Proteins at Low Temperatures

Owen Fennema, EDITOR
University of Wisconsin–Madison

Based on a symposium sponsored
by the Division of
Agriculture and Food Chemistry
at the 175th Meeting of the
American Chemical Society,
Anaheim, California,
March 12–17, 1978.

ADVANCES IN CHEMISTRY SERIES **180**

AMERICAN CHEMICAL SOCIETY
WASHINGTON, D. C. 1979

Library of Congress CIP Data

Proteins at low temperatures.
(Advances in chemistry series; 180 ISSN 0065-2393)

Includes bibliographies and index.

1. Proteins—Congresses. 2. Cold—Physiological effect—Congresses. 3. Proteins—Denaturation—Congresses.

I. Fennema, Owen R. II. American Chemical Society. Division of Agricultural and Food Chemistry. III. American Chemical Society.

QD1.A355 no. 180 [QP551] 540'.8s [574.1'9245]
ISBN 0-8412-0484-5 ADCSAJ 180 1–254 1979
79-16561

PRINTED IN THE UNITED STATES OF AMERICA

Advances in Chemistry Series

FOREWORD

ADVANCES IN CHEMISTRY SERIES was founded in 1949 by the American Chemical Society as an outlet for symposia and collections of data in special areas of topical interest that could not be accommodated in the Society's journals. It provides a medium for symposia that would otherwise be fragmented, their papers distributed among several journals or not published at all. Papers are reviewed critically according to ACS editorial standards and receive the careful attention and processing characteristic of ACS publications. Volumes in the ADVANCES IN CHEMISTRY SERIES maintain the integrity of the symposia on which they are based; however, verbatim reproductions of previously published papers are not accepted. Papers may include reports of research as well as reviews since symposia may embrace both types of presentation.

CONTENTS

PREFACE

It is well known that proteins are major and vital components of living matter and of human diets, and that researchers, in recognition of this fact, have studied them so intensively that research libraries are crammed with monographs and journals containing results of these investigations. It is therefore strange that comparatively little effort has been devoted to elucidating the critical role that proteins play in determining responses of biological matter and foods to low temperatures. This subject, which obviously is considered of secondary importance by most protein researchers, is deserving of considerably more attention than it has received since it is generally believed that proteins are involved in major but poorly understood ways with the behavior of living matter and foods in the following low-temperature situations:

1) The ability of some animals to hibernate and others to acclimate when exposed to low nonfreezing conditions.

2) The ability of some plants to "winter harden" and thereby tolerate freezing conditions.

3) The ability of some fish to avoid freezing in cold water.

4) The ability of some microbial cultures and other small biological specimens to survive freezing, frozen storage, and thawing.

5) The inability of humans to tolerate low nonfreezing temperatures.

6) The inability of whole organs, large biological specimens, and whole animals to survive frozen storage. (Rectifying this inability would have immeasurable benefits to the human race.)

7) The inability of plant and animal food tissues to withstand freezing and frozen storage without undergoing detrimental changes such as a decrease in water-holding capacity and alterations in texture (toughening of fish muscle, softening of delicate plant tissues).

The purpose of this volume is to draw together summaries of major research studies that have dealt with the behavior of proteins at low temperatures, and to thereby stimulate additional research in this area. The book is organized to provide an "overview" chapter followed by chapters on low temperature behavior of proteins in simple systems, in plant tissues, and in animal tissues.

Department of Food Science
University of Wisconsin–Madison
Madison, Wisconsin 53706
February 8,1979

OWEN FENNEMA

1

Protein Alterations at Low Temperatures: An Overview

GEORGE TABORSKY

Section of Biochemistry and Molecular Biology, Department of Biological Sciences, University of California, Santa Barbara, CA 93106

Physical and chemical alterations of protein structure and function, caused by the exposure of aqueous protein solutions to low temperatures, have been explored and explained by various investigators in terms of a broad range of hypotheses. This survey attempts to give a mostly retrospective overview of this range, emphasizing effects of low temperature and of freezing on hydrophobic interactions, hydrogen bonding, and interactions of the protein with the solvent, other solution components, the ice, and the ice–liquid interface. The effects of low temperatures per se *are considered, followed by a discussion of those effects on protein structure and reactivity which are associated with ice formation and its consequences such as the freezing out of solutes. The effects of the admixture of organic cosolvents to aqueous protein solutions are noted with reference to cryoenzymological studies and the prevention of freezing injuries. Experimental and interpretive approaches and results of these approaches are illustrated by examples drawn selectively from the literature.*

Water is, of course, the primal matter. This was as clear to the ancient philosophers of Greece and India as it is clear to modern students of their own internal and external environments. The physical states of water, and the process and consequences of its phase transitions, have always been near the center of the overall interest in water. The physical chemist is attracted by the "anomalies" associated with these states and

0-8412-0484-5/79/33-180-001$6.50/0

phase transitions. The geologist is drawn by the abundance of the aqueous cover of the terrestrial surface and the dynamics of this cover in its various states of aggregation. For the biologist, water is a substance that is intimately and pervasively involved in the maintenance of the vital integrity of biological structures and in the expression of many of their vital functions. Indeed, the modern biologist clearly recognizes the essential reciprocity of this involvement: the structure of water is linked in a mutual relationship with the order and organization of biomatter.

These varied and fundamental vantage points would suffice by themselves to account for interest in water and its physical states. But interest in the properties of water and of aqueous systems, in all states of aggregation, has been sustained also by demands of practical applications in fields as diverse as meteorology, medicine, agriculture, technology, and commerce—to name a few.

Despite the long history of physical and biological interest in freezing, an effective joining of physical-chemical and biological insights is a recent achievement. The editor of a landmark review of the many facets of "cryobiology" has duly noted that a significant confluence of physical and biological thought on freezing barely existed prior to 1940, and even in the 1960s only a handful of investigators were effectively engaged in the study of cryobiology at the fundamental level (*1*).

The last two decades brought welcome change in this regard. Aqueous biological systems are being studied at low temperatures on a broad front and in a manner that makes integration of diverse disciplinary insights increasingly promising. The theme of this symposium reaches the core of these efforts in that the behavior of proteins (and, by inference, other macromolecules) at low temperatures must be viewed as the key to any real understanding of cryobiology. Obviously, the book on freezing of biological systems is not about to be closed. But it seems equally certain that on-going efforts—many of which are reported in this volume—are impressive and highly encouraging for the future.

The role assigned to this paper is to provide an overview of proteins at low temperatures. This is a very broad assignment. Within available space, response to this assignment must be unavoidably selective and cursory. But the theme of this charge is largely coextensive with the theme of the symposium. Its special aspects will be explored in more critically searching fashion throughout the bulk of this volume. I shall not encroach on those topics that will get expert coverage in the specialized chapters to follow. Instead, I shall attempt to give a summary account that will be largely retrospective. It seems that such an effort at the outset could be of value to the extent that it provides a fair view of the range over which problems of cryobiology have been approached and thought about.

As subsequent chapters will undoubtedly show, the intellectual and experimental approaches taken to cryobiology in the relatively recent past have clearly advanced our understanding of freezing phenomena but, the advances of the last decade or two notwithstanding, many of the problems continue to challenge the cryobiologist. It would be premature, it seems, to attempt to draw an integrated picture of the state and dynamics of proteins in low temperature systems. Too many features of the picture remain insufficiently clear for a precise definition of their relative place within the whole. However, the problems can be addressed one at a time while keeping in mind that fragmentary and oversimplified concepts must eventually merge into an integrated description if the reality of freezing phenomena is not to be misrepresented.

Conceptual Approaches to the Study of Proteins at Low Temperatures

It is by no means certain that a given freezing experiment, designed or interpreted in a particular conceptual setting, is optimally effective in advancing our understanding of freezing phenomena. The freezing literature is not free of ambiguities or controversies that stem from hypotheses advanced on a too narrow or even mistaken basis. The identification of the proper basis on which to plan or evaluate experiments represents the key to progress. It is likely that all of the conceptually significant features of the freezing process have not yet been recognized. Thus, important factors may elude proper experimental control. Also, we may find ourselves attracted by reasonable notions that turn out to be not amenable to definitive experimental tests. It seems best to try to deal with these reservations by keeping them in focus.

With these reservations in mind, it seems safe and helpful to take the pragmatic approach and proceed with organization of the subject matter in terms of the most frequently employed working hypotheses. These tend to be rooted in particular features of the freezing process or the frozen state. I propose to take these under consideration in the approximate order of their experimental or conceptual complexity.

The simplest conceptual framework for a low temperature or freezing study is the assumption that any observed difference between protein structure or behavior at "normal" and at "low" temperature can be accounted for in terms of the temperature difference alone. This hypothesis focuses attention on a cold protein solution that may be homogeneous or in contact with frozen solvent. If there is ice present, it is considered of no particular significance. We can expand this view next by recognizing that freezing has certain consequences in addition to a change in temperature. Furthermore, we can take the tack that the key to protein alterations is the interaction of the protein with the frozen framework in which it is embedded. And lastly, we can direct our attention at the freezing process itself rather than the frozen system in its equilibrium or quasi-equilibrium state. In this case, we would take cognizance of the possibility that alterations of the protein occur where the dynamic events of the phase change take place, namely at or near the ice–liquid interface. These events may be linked directly to the production of structural changes or they may potentiate such changes, to be expressed later through altered protein behavior.

Proteins in Cold Solutions

For our purposes, a protein solution is "cold" whenever its temperature is near or below the freezing point of water. The freezing point can be readily manipulated by adding low molecular weight solutes. Thus, the temperature range over which the solution can be studied in the absence of ice can be appreciably extended—by 20 or more degrees below zero. The lower limit of this manipulation is defined by the eutectic temperature(s) of the particular system (*2*). More drastic depression of the freezing point (to $-100°$ or even lower) can be attained upon the admixture of certain organic solvents. The use of such "cryosolvents" for the particular purpose of protecting cells against freezing dates back about thirty years, when glycerol was first employed as a cryoprotectant (*3*). Mostly for the same purpose, many other organic solvents have been used and investigated since then (*4*). Their use with particular emphasis on consequences at the molecular level, especially on proteins, is of relatively recent vintage (*5, 6*).

Protein Conformation and Low Temperature: General Considerations. Narrowing the focus to just the protein solute and its immediate liquid environment is admittedly an oversimplification. However, taking the narrow view, if only for the moment, has its advantage. It will emphasize that whatever else may occur, the protein will always adjust its conformation during a temperature change in accord with the temperature dependence of those forces that maintain the protein's confirmation. Although a definitive analysis is beyond the capabilities provided by theorists, a qualitatively reasonable and quantitatively emerging picture is at hand and may be useful to describe the effects of low temperatures on protein conformation. In our discussion of this picture, we shall rely heavily on a recent pertinent review (*7*).

It is generally accepted that the principal contributor to the stability of the "native" conformation of a protein is the tendency of its nonpolar groups to avoid contact with water—that is, to engage in hydrophobic interactions. Based on model compounds and on current theories of water structure, the process is viewed as largely entropy-driven. The unfavorable low entropy of "ordered" water, which would surround nonpolar groups when these are exposed to the aqueous medium, is avoided because the nonpolar groups fold inward. This is accompanied by an entropy-increasing release of ordered water to the relatively disordered state of the bulk solvent. Considering that bulk water itself becomes more extensively hydrogen-bonded and thus more highly ordered as the temperature is lowered, the entropic advantage of folding nonpolar groups inward is lessened at low temperatures. In this case, stabilization by hydrophobic interactions would be expected to decrease. [It is generally unquestioned that "hydrophobic interactions" are of major importance with regard to the conformational integrity of proteins and that these interactions suffer when a protein is "cold-denatured." Many experimental data are supportive of this view (see, for example, *8, 9*). This support not withstanding, certain experimental parameters of protein denaturation —particularly the associated, unexpectedly small volume changes (*10, 11*)—have caused questions to be raised concerning the strict applicability of models in terms of which hydrophobic interactions, maintaining proteins in a folded state, had frequently been accounted for. This problem is given critical attention in a number of recent reviews (see, for example, *7, 12, 13, 14, 15*).]

A strongly contrasting behavior may be expected, at first sight, with hydrogen bonds that contribute to protein stabilization. Low temperatures should enhance hydrogen bond formation in view of the negative change in enthalpy that characterizes this process. We would expect hydrogen bonds to become stronger as the temperature drops. Such strengthening should apply to intramolecular bonds (between donor and acceptor groups of the protein) as well as to intermolecular bonds (between protein groups and water molecules). The enhanced strength, particularly of those intramolecular bonds in the protein's interior, might be expected to enhance conformational stability—or even to alter the conformation by upsetting the overall balance of all forces that maintain conformation. There appears, however, to be little likelihood that additional hydrogen bonds will form in the interior of the protein during cooling since most of the possible bonds would be expected to exist already at the higher, "normal" temperatures. [The view that most of the possible hydrogen bonds already exist in the "normal" protein may require qualification by the postulate (*12*) that the demonstrated accessibility of the protein interior to water (and to H^+ or OH^-) is due to structural "defects," existing because of suboptimal hydrogen bonding of peptide groups. According to this view, *some* possibility remains in a "normal" protein for the acquisition of additional hydrogen bonds as the temperature is lowered.]

There is little that we can say about electrostatic forces in connection with low temperature effects. Undoubtedly, the charged groups of a protein interact and thereby contribute to conformational stability. However, the magnitude and direction of these contributions is beyond our ability to assess since it would require information about their precise location on the protein surface and about the effective dielectric constant on which the force of these interactions depends. Such information is usually lacking. Thus, even though temperature would be expected to influence charge density (because acid dissociation equilibria are temperature dependent) and the dielectric constant (whatever its value), our ability to predict is severely circumscribed. Effects may be significant but their magnitude cannot be assessed nor is it possible to determine whether such effects would strengthen, weaken, or alter the protein's conformation in a temperature dependent manner.

In summary, we can say that low temperatures would be expected to alter protein conformations by favoring the formation of and strengthening hydrogen bonds and diminishing the importance of hydrophobic interactions. The effect of low temperatures on electrostatic interactions in proteins is outside our pale. We should also recall the earlier stipulation that whatever consequences the cold or frozen state might have on proteins per se, these consequences must be integrated, in the final analysis, with the consequences of cooling or freezing on nonprotein components of the system. The present "minimal" hypothesis as well as those which are yet to be considered, taken individually, provide only a limited view of low temperature and freezing phenomena.

Experimental Observations of Putative Conformational Changes Ascribed or Ascribable to Low Temperature Effects. Some of the path-breaking studies of the effects of near-zero or subzero temperatures on several enzyme systems were interpreted on the basis of assumed, temperature-dependent shifts in enzyme conformation. These conformational changes were presumed to relate directly to the degree of intramolecular hydrogen bonding (*16, 17, 18, 19, 20*). While the physical-chemical description of the observations is reasonable and straightforward (*cf. 20*), the hypothesis must be viewed with reservations, not only because it was proposed at a time when little was known about mechanisms and forces by which protein conformation is stabilized, but also because the evidence in support of the hypothesis was inferred rather than direct. Nevertheless, these results should be noted: they were the first results of a first attempt at quantitative insight into a highly refractory problem and they may still be instructive. Table I illustrates the nature of this insight. The kinetic data obtained with three enzymes fitted biphasic plots of log k vs. $1/T$. These plots yielded values for the apparent Arrhenius activation energies, which showed a major increase when the temperature was lowered below the point at which the aqueous system began to freeze. Similar biphasic Arrhenius plots were subsequently reported for other enzymes (e.g., alkaline phosphatase and peroxidase)

Table I. Arrhenius Activation Energies of Some "Frozen" Enzyme Reactions[a]

	Arrhenius Activation Energy (kcal/mol)	
Enzyme	*Above 0°*	*Below 0°*
Lipase	7.6	37.0
Invertase	11.1	60.0
Trypsin	15.4	65.0

[a] All data were taken from Sizer and Josephson (*16*).

(*18*). These results were interpreted to mean that low temperatures favored the formation of enzyme molecules with abundant hydrogen bonds (intramolecular) and caused, thereby, a decrease in enzyme activity. [Although, as noted, this interpretation predates the introduction of the concept of hydrophobic interactions and perforce ignores the putative importance attributed to these interactions as the major conformation-maintaining force, the notion of enhanced hydrogen bonding at low temperatures should not be simply dismissed as outdated. The recently postulated "structure defects" attributed to incompletely hydrogen-bonded structures in the native protein interior (see Reference *12*) are, in principle, fully capable of completion at lowered temperatures, causing conformational rearrangements on a scale that may be significant.] It should be emphasized that this hypothesis was even then regarded as overly simplistic in that it probably overlooked a whole range of other consequences of exposure to low temperature. Recognized possibilities included: 1) a viscosity increase, 2) changes in intermolecular interactions involving enzyme, substrate, solvent, and buffer ions, and 3) shifts in ionization equilibria of any of these components. Thus, the stage was set for future investigations of a more searching nature.

Once the importance of hydrophobic interactions became generally evident, it became clear that these interactions might undergo important changes whenever proteins are exposed to low temperatures. This view dominates reports on multimeric enzyme systems, micelle-like aggregates, and membranous structures. I will refer to several examples from the relatively recent literature for purposes of illustration.

Prompted by the striking initial observation that multiple enzyme forms are produced upon freezing solutions of lactic dehydrogenase (*21*), the properties of several other enzymes were explored under freezing conditions (*22*). These studies may be viewed as the take-off point for many studies that gave careful attention to the diverse parameters that can affect the outcome of a freezing experiment. It produced the tentative generalization that subunit dissociation may be the primary response of enzymes to freezing.

By noting the behavior of casein micelles of genetically varied polymorphic composition (*23*), one can readily sense that freezing may have only subtle effects on the molecular structures of macromolecular complexes. The polymorphism of casein stems from minor variations in the primary structure of casein subunits. It was found that freezing produced different changes in micelle stability depending on the genetic variant being studied. For example, in the case of one particular variant complex, its stability (nonfrozen) exceeded that of the average casein micelle, but upon freezing this same variant complex became less stable than the average micelle. All of this presumably occurred because of a few amino acid substitutions. It is almost necessary to assume that the change in stability of the micelle reflects some rearrangement in the manner in which subunits interact. An independent observation (*24*) tends to support this view. Long frozen storage was found to cause distortion and, ultimately, depolymerization of the micelle.

The egg provides another system of interest because of the role that lipoprotein complexes play in the yolk structure. Egg yolk gels on freezing. There is broad unanimity in the literature, based on a wealth of experimental data (*25–30*), that this response to freezing involves lipoprotein aggregation. Replacement of native hydrophobic interactions by artificial ones may occur because the native arrangement may be disrupted by freezing-induced changes in the lipoprotein's aqueous environment. Importance has also been ascribed to the fatty acid spectrum of yolk lipids [influencing the temperature of thermal transitions (*26*)] and to the salt content [affecting the integrity of the yolk granules of which the yolk lipoprotein is a major component (*25, 27, 29*)] The formation of additional hydrogen bonds during freezing has also been suggested (*29*).

Hydrophobic interactions are, of course, crucial to biomembrane integrity. Membranes, therefore, should exhibit vulnerability to freezing, and they do. It seems, however, that injuries need not primarily concern the *protein* chemist. For example, the fluorescence of the protein of erythrocyte membranes undergoes a shift on freezing. This is indicative of a change in the protein environment, most likely involving membrane lipids (*31*). Proteins are apparently not released (*32*). Nevertheless, membrane-bound proteins do not escape the effects of low temperature altogether. It has been shown, for example, that the circular dichroism exhibited by the erythrocyte membrane (due, presumably, to its protein components) undergoes significant changes as the temperature of the membrane suspension is varied over a wide range, down to nearly 0°C (*33*). (Sharp, cooperative transitions reflecting denaturation were shown to occur only at temperatures above 40°C under the conditions of these experiments.)

The mitochondrial membrane also suffers alterations of its lipid content during freezing. For example, freezing alters ketoglutarate dehydrogenase activity and this change may result from the inhibitory action of free fatty acids that have been generated, presumably, by freezing-induced activity of phospholipases (*34*). In another study, the "antiradical activity" of lipids from various frozen and thawed liver organelles was shown to diminish (*35*). Studies with frozen brain organelles lend additional strength to the view that protein damage is not, perhaps, the primary, functionally destructive event during freezing of membrane systems. In these studies, some redistribution of "marker" enzymes among subcellular fractions was found but loss of enzyme activity was minimal, suggesting that the membrane binding sites of these proteins were damaged rather than the proteins themselves (36).

Before I turn to more complex matters, I would like to emphasize that many of the putative conformational or other changes mentioned above occurred in a frozen rather than just a cold environment. Thus, the changes observed probably involve more than a simple exposure to low temperatures. However, in some instances the effects of low temperature per se may predominate, and this simple hypothesis always merits consideration.

Frozen Aqueous Solutions: Concentration Effects. As solute is rejected by the growing ice and as its concentration increases in the shrinking liquid phase, the temperature drops toward the eutectic point, where the entire system approaches complete solidification. Of major importance is the fact that a highly concentrated liquid phase can persist indefinitely at any point above the eutectic temperature.

This type of "freezing-out" has long been known and has been exploited in industrial settings. But, in a research setting, attention was focussed on it only recently when the cryochemist became concerned with "anomalous" reaction kinetics in "frozen" systems, and when the cryobiologist began looking into the causes of freezing injury. An illustration of such studies, as they pertain to proteins, is now in order.

The typical protein solution contains a variety of low molecular weight solutes (e.g. substrates, cofactors, buffer salts). Such systems should exhibit several "eutectic" points. During freezing, the concentrations of various molecular species, because they have different eutectic points, will change relative to each other. These changes will continue until the final eutectic temperature has been attained. For example, relatively simple aqueous systems composed only of sodium and potassium salts of phosphoric acid were shown to have eleven eutectic points associated with them (see Table II). It is important from a protein-oriented point of view that the pH values and ionic strengths of these eutectic solutions range widely within their freezing range between 0°

Table II. Eutectic Solutions of Phosphate Salts[a]

Solute(s)	*Ionic Strength (M)*	*Freezing Point (°C)*	*pH[b]*
Na_2HPO_4	0.33	−0.5	8.9
KH_2PO_4	0.92	−2.7	4.1
$Na_2HPO_4 + KH_2PO_4$	2.50	−4.3	5.5
NaH_2PO_4	3.42	−9.7	3.3
$NaH_2PO_4 + Na_2HPO_4$	3.60	−9.9	3.6
$NaH_2PO_4 + KH_2PO_4$	3.84	−11.2	3.4
$NaH_2PO_4 + Na_2HPO_4 + KH_2PO_4$	4.10	−11.4	3.6
K_2HPO_4	8.55	−13.7	9.3
$Na_2HPO_4 + K_2HPO_4$	9.00	−14.4	9.3
$KH_2PO_4 + K_2HPO_4$	9.40	−16.7	7.5
$Na_2HPO_4 + KH_2PO_4 + K_2HPO_4$	9.72	−17.2	7.5

[a] All data were taken from Van den Berg and Rose (*37*); the ionic strength values were calculated from concentration data given in the reference.
[b] pH measurements were made at 25° after separation of the unfrozen liquid from the solid phase.

and about − 17°. Thus, a dissolved protein's structural integrity can be expected to be challenged by these variations. Milk, for example, in which the proteins can experience freezing-induced changes, will show a decline in pH on slow freezing from 6.6 to 5.8. Fast freezing causes no immediate pH change, but a low pH develops upon "frozen" storage—presumably because equilibrium is slowly established between liquid and solid phases (*38*). Freezing-induced changes in pH occur in many solutions and foods (*39*) and it is to such pH variations that specific enzymic activity losses upon freezing have been attributed (*40*). Such freezing-induced pH effects can be counteracted by the addition of "cryoprotectants" such as glycerol or dimethylsulfoxide, which are effective because the selective precipitation of buffer salts will occur only at lower temperatures in their presence (*41*). There are numerous reports of protein alterations attributed to other concentration changes as well. Denaturation of chymotrypsin (*42*), activation of a phosphodiesterase (*43*), and disruption of yolk granules (*29*) have been ascribed, at least in part, to increases in salt concentration upon freezing. (Parenthetically, we may note some technically interesting developments. An apparatus has been described that permits the freeze concentration of proteins themselves, on a multiliter scale. Ten-fold increases in concentration can be obtained by this means (*44*). Also in the practical domain is a recent report on the use of pH indicators in frozen solutions (*45*). Finally, we may take note of an experiment that is indicative of the selectivity of the freezing-out mechanism. Upon freezing a mixture of light and heavy water (deuterated), an isotope enrichment amounting to 2% could be achieved (*46*).)

In general, it is clear that freezing can cause concentration of solutes to change by orders of magnitude. Changes in concentration of this degree should be able, potentially, to cause major alterations in proteins. Furthermore, such alterations need not be confined to noncovalent changes. Substances capable of reacting covalently with a protein may undergo such reactions when pushed by the mass action of their increased concentrations.

Reactions in Frozen Aqueous Systems. Particularly during the sixties, there was a wave of concern about rates and mechanisms of chemical reactions that occur in the "frozen state." Earlier observations of chemical changes were largely accidental and frequently held to be only of nuisance value. The one major exception concerned cryobiological inquiries into the causes of freezing injury. In these cases, however, the complexity of the material effectively masked the chemical roots of the mechanism. Even though most studies of "frozen" chemical reactions have involved low molecular weight reactants, they merit consideration because their conceptual framework is certainly applicable to complex biomolecules or other biomatter.

Much of the interest in these reactions stemmed from the attractiveness of the possibility that special properties of ice may be responsible for novel reaction mechanisms. For instance, proton mobility is higher in ice than in liquid water by one or two orders of magnitude and the effective dielectric constant in ice is lower than that of water by about an order of magnitude (*47*). Under such conditions, mechanisms involving proton transport or charge interactions may change, even qualitatively. Changes in membrane permeability, for instance, may reflect changes in dielectric properties of membranes resulting from altered proton behavior at low temperature (*48*). An entirely different speculation has been advanced regarding the possibility of reciprocal effects between ice-like structures and the stabilization of electronically excited states of biomolecules, thus giving rise to a hypothetical, novel energy transduction mechanism (*49*).

A major point was made after a critical review of numerous reports on reaction rates in frozen systems (*50*). Kinetic–mechanistic "surprises" in "frozen" systems may not require exceptional hypotheses. Concentration effects may account for them. Even if a system appears to be completely solidified and, therefore, not amenable to analysis in terms of unfrozen liquid "puddles" (*51*), a liquid phase should still be considered a possibility. A case in point is provided by the efficient electron transfer observed between ferrous and ferric ions in an aqueous system frozen below its putative eutectic point (*52*). This seemed to require an ice structure in order to bridge the distance between reactants that were calculated to be too far apart for significant reactivity. However, it was pointed out that the assumed eutectic point was based only on the major solute component, whereas minor components—the reactants, in this case —could have lowered the effective eutectic point below the reaction temperature, making the assumption of a special mechanism superfluous (*50*).

Elucidation of mechanisms of frozen reactions will be hindered if the coexistence of liquid and solid phases is viewed necessarily as a thermodynamically defined phenomenon, implying stable phases and concentrations. Frozen reactions could occur in a liquid phase that is only of transient significance. In the trivial case, "puddles" may exist simply because freezing requires finite time. In other cases, more sophisticated insight into the dynamics of the freezing process may be needed. For example, transient puddles may exist in a variable manner depending on the rate of cooling. Unfortunately, the required insight is elusive because the process is too complex. A highly relevant, thoughtful analysis of the distinction between cooling kinetics (i.e. temperature change) and freezing kinetics (i.e. solidification) is available (*53*). An important feature of this distinction is that freezing velocity depends on the ability of the system to dissipate the heat of crystallization. This requires time, during which the supercooled liquid must serve as a temporary heat sink while it exists entrapped within a "spongy" ice structure (*54*). These liquid pools can be the medium in which chemical reactions take place under conditions that are largely indeterminate. For example, the uniformity of this solution in terms of concentrations and temperature cannot be taken for granted, and any concentration and thermal gradients are, for all practical purposes, beyond precise, quantitative description. In addition, relatively little control can be exercised over the rate of heat withdrawal and this determines the size of the liquid pools and how closely the ratios of the solid and liquid phases approach equilibrium. There also may be irregular thermal conductivity gradients in the system which, together with features of gross geometry (surface/volume ratio and shape), will have a major impact on the manner and speed with which the heat of crystallization is finally taken up by the cooling source. This is not an encouraging picture but it is a realistic outline of the problems that cannot be ignored—at least not in those cases in which rapid, concentration-driven changes may occur in liquid lacunae that exist (permanently or transiently) within the growing ice.

There are not many reports of covalent, freezing-induced changes that are relevant to protein chemistry. A few examples can, however, be cited. A freezing-dependent enhancement of addition reactions involving the formyl groups of heme has been noted (*55*). Presumably, a concentration effect operates here, expressed through a pH shift toward the reaction optimum. Of perhaps greater significance is the hypothesis that freezing injury of proteins may come about through the formation of intermolecular disulfide bonds (by air oxidation or by disulfide interchange), promoted by the freezing-induced unfolding and concentration of proteins (*56*). For example, a set of enzymes known to be inactivated by freezing was recognized some time ago as being of the sulfhydryl variety. Cryoprotection was afforded for some of these enzymes by the addition of mercaptoethanol (*57*). Denaturation of myofibrillar proteins upon frozen storage was attributed to the presumed oxidation of sulfhydryl groups (which were shown to undergo a loss as a consequence of freezing) (*58*). More recently, oxidation of a sulfhydryl group in actin was demonstrated to occur on freezing, resulting in formation of a dimer (*59*). The freezing-induced inactivation could be reversed, or prevented, if a reducing agent was added. However, it seems that using mercaptans as cryoprotectants would be counterproductive whenever the protein depends for its integrity on the intactness of its disulfide bridges. When mercaptans are desirable, their reactive effectiveness can usually be enhanced by the concentration effect associated with freezing. For example, significant alterations of peanut proteins can be produced by reducing agents, and their addition prior to frozen storage of such protein solutions can promote alterations that are dependent on reducing agents (*60*). Whenever proteins are more stable in a reducing environment, this approach may be useful.

A final comment should be made. Protein structure is maintained by many "weak" interactions. During freezing, we must be prepared for the possibility that conformation-maintaining forces may be driven to seek a new balance, via a conformational adjustment. This could occur in response to interaction of the protein with "ligands" present in the solution, especially if these ligands mimic the role of protein groups involved in weak interactions in the native structure. These ligands may compete with intramolecularly interacting groups. The freezing-induced concentration of such ligands may result in significant conformation changes even with ligands that would, under "normal," more dilute conditions bind too weakly to have an effect. (For a relevant discussion, see Ref. *61*).

Cryoenzymology (*62*). Recent low temperature studies of enzymes merit separate consideration. They will be given special emphasis in a later chapter of this volume. These investigations have opened a new window through which we can gain insights into the dynamics of enzyme action with a resolving power unattainable, or attainable only with difficulty, at "normal" temperatures. This capability hinges on the use of cryosolvents that permit attainment of very low temperatures while avoiding the complications of the formation and presence of a solid phase. The topic was recently reviewed and the results of cryoenzymological studies were integrated with results gained from investigations of enzyme dynamics in the crystalline state (*63, 64*). The major attraction of this approach lies in its potential for yielding information about enzymic processes under conditions where otherwise inaccessible intermediates accumulate and become amenable to characterization. Obviously, artifacts of doubtful significance to the true biological mechanism could be expected to form in the presence of organic solvents. However, careful assessment of such undesired effects is possible and there is already an appreciable number of enzymic systems for which mechanistically important features of the catalytic process have been identified (*63*).

The possibility of preparing protein crystals in equilibrium with cryosolvents (*64*) mutually enhances the value of crystal analysis by X-ray diffraction at low temperatures and of the cryoenzymological approach. By this means it has become possible to compare reactive intermediate forms trapped in the crystalline state and in solution, respectively (*63*).

Cryosolvents and Protection Against Freezing Injury. Cryosolvents can assist the biologist whose goal is to protect biological systems from freezing injury and thereby enable long-term preservation of functionally intact tissues and cells. Protective cryosolvents may not alter the proteins in a particularly significant manner, rather they help assure that the protein's integrity remains essentially unaffected. The effects of cryoprotectants will not be completely understood until it becomes more fully clear what their effects may be on the structure of the aqueous medium on which, in turn, the conformational features of the protein component so intimately depend. In addition, the eventual explanation of cryoprotection in molecular terms will require more insight into the nature of the direct interactions between the protein and the protective solvent component. (Relevant discussions are provided, for example, in References *7, 65, 66, 67*). Another aspect of the effectiveness of cryoprotectants is the fact that they penetrate membranes with ease and at least part of their function may be to extend the range of temperatures over which the cell contents can exist in a liquid state and to avoid development of deleteriously high concentrations of cellular constituents during freezing. Selection of these protective cryosolvents is likely to be based increasingly on information obtained from cryoenzymological studies that reveal fundamental characteristics of cryosolvent–protein interactions as a "by-product."

The matter of cryoprotection, with clinical applications in mind, will undoubtedly become better understood with time. Obviously, measures of cell or tissue viability must be thought of with more subtle and stringent criteria in mind than those normally used in most studies of tissue preservation. Assessing viability in terms of one or a few biochemical parameters rather than in terms of a whole sweep of physiological characteristics of a given tissue spells the difference between adequacy for a limited purpose, such as in vitro experiments, and suitability for, say, tissue transplantation. The complexity of this problem will increase further when consideration is given to potential interactions between cryoprotectants and other pharmacologically active agents and to the metabolic state of the tissue. Although the art of cryoprotection is not yet at a point where these considerations can be effectively dealt with, progress is being made (*68*).

Proteins in the Presence of Ice

A critical relationship exists between water and protein (*69*). I have found it convenient to deal earlier with one manifestation of this relationship, namely hydrophobic interactions. It seems obvious, however, that this relationship should be affected by any significant alterations in the state of the aqueous component, regardless of whether the water is attempting to avoid contact with nonpolar groups or is attracted to polar groups. Although this assertion is obviously true, difficulties arise when a rigorous analysis is attempted. The problem begins with our limited understanding of water itself—more than one model of water structure can be fitted to available data (*70*). Furthermore, theories of water–protein interactions are incomplete and tentative (*14*). Thus, an accurate description of water–protein interactions in complex biological systems is not yet possible (*71*). [It is noteworthy that the authors of a recent review of the role of solvent interactions in protein conformation felt compelled to note that the "veritable jungle of data and hypotheses" produced by many investigators "working in closely allied fields" but with a "general lack of coordination" might invite the consideration of "a more purposeful, joint attack on some of the problems" which "may well be timely and yield valuable dividends" (*13*).]

Against this somewhat forbidding background, I will engage in a cursory review of some of the problems associated with protein–water–ice systems wherein *ice* is suggested, or acknowledged, to have a role in protein alterations. At this point, attention will be given to the effects of freezing on hydration of proteins. Such effects must lead to alterations of protein structure if the hydration shell plays an integral role in maintenance of protein structure. (Hydrophobic interactions have been dealt with already in the context of protein behavior in cold solutions. I will simply add here that it is clearly unavoidable that diminishing the liquid environment of the protein by freezing will remove the raison d'ètre of these interactions and will therefore weaken this form of structure stabilization.) Our discussion of freezing effects on protein hydration can be brief since much of the pertinent information and many of the relevant ideas have been thoughtfully reviewed quite recently (*15, 67*) and will be dealt with later in this volume.

Protein Dehydration Upon Freezing. Most of the current models of protein hydration postulate the existence of several classes of water molecules that are distinguished by different degrees of "structure," mobility, and strength of interaction with the protein. Table III provides an "orders-of-magnitude"-type summary of these classes. "Site-bound" water consists of water molecules bound individually and stoichiometrically to specific, typically charged or polar sites. This water is bound with relatively great strength and exhibits greatly reduced mobility compared to bulk water. "Surface" water is water adjacent to the protein structure, held with moderate strength, intermolecularly hydrogen-bonded, and exhibiting hindered mobility. While "water of hydration" must still be considered as an operationally defined term (different "classes" of water being identified differently, depending on the particular experimental method chosen to reveal them), most experimental determinations of protein hydration are likely to deal, more or less, with these two categories

Table III. Classes of Water Molecules in Protein "Hydration Shells"[a]

Class	*Binding*	*Rotational Relaxation Time (sec)*	*Molecules of Water per Residue*	*Degree of Hydration (g/g)*	*Freezability*
Protein-bound					
site-bound	strong	10^{-6}	0.1–0.5	0.02–0.1	−
surface	medium	10^{-9}	1–3	0.2 –0.6	−
bulk-like	loose	10^{-11}	—	—	+
Monolayer	—	—	4–7	0.7 –1.2	—
Protein-free					
liquid	—	10^{-11}	—	—	+
ice	—	10^{-5}	—	—	—

[a] Information collected in this table is based on data from Kuntz and Kauzmann (*67*) and from Richards (*15*). Numerical values given here represent approximations based on numerous proteins and diverse experimental techniques. The class designated as Monolayer is included for purposes of comparison with "Surface" water in particular. The latter is, presumably, an actual layer of uneven thickness that is variable from protein to protein. The former is an idealized concept of a uniform layer, one molecule in thickness, covering the protein surface. "Freezability" denotes whether or not a particular class of water is capable of transition into normal ice.

of water. It is noteworthy for our purposes that "site-bound water" or "surface water" is not capable of forming normal ice—it is more mobile than water molecules in ice, even at low temperatures, but it is less mobile than water in the bulk solvent at the same temperature (*67*). Although water of hydration appears to be "structured," it is not "ice-like" (*72*). The third class, designated as "bulk-like," is in most ways indistinguishable from bulk solvent. Its existence is acknowledged only because it may contribute to some of the hydrodynamic properties of the protein. Presumably, the demarcation between "free" solvent water and "hydration" water is not a sharp one. The "bulk-like" water is envisaged as representing the transition from one to the other.

It should be stressed that this classification is an oversimplification. The properties of water of hydration most likely change more gradually that is implied by "classes" (*67, 73*). Nevertheless, they are useful in that they facilitate a reasonable conceptualization of many of the experimental data that bear on protein hydration. But it is prudent to keep in mind that these classes probably represent averaged segments of a continuum of water structures. In any case, whatever may be its precise definition, the reality of the hydration shell is profusely documented (*67*). It must be viewed as an integral part of the protein, the functionally significant entity. Indeed, protein hydration is of critical importance to the protein's functional integrity insofar as this function is an inherent molecular property—a specific manifestation of protein structure. The reality of the hydration shell appears also from the vantage point of a reactive ligand to which the hydration shell seems to represent a physical barrier to be surmounted before the functionally significant, direct interaction between protein and ligand can be accomplished. This has been shown, for example, with respect to the interaction between myoglobin and oxygen, or carbon dioxide, where the initial association between ligand and protein is seen as a process that is slower than diffusion-limited and is attributed to ligand entry from bulk solvent through the hydration shell, a process noticeable down to temperatures as low as about −60° (in glycerol–water mixtures) (*74*).

Interpreting data relating to the effects of freezing on protein integrity and protein hydration is difficult since most studies provide inferences rather than proof. The closest to "proof" are those studies involving the attributes of dried proteins and their relation to proteins in frozen systems. What seems necessary is to show that disturbance of the hydration shell, perhaps by ice crystals or by more subtle changes in the structure of water of hydration, has notable consequences on protein structure. That the drying of proteins has major consequences of this sort has been shown. This is, of course, as we would expect. If the more intimately bound layers of the hydration shell are removed, the original protein conformation would have to accommodate large stresses produced by the structural voids. Transconformation or denaturation would be the appropriate and accessible response of the protein to such stresses (*67*).

Some examples may serve to support this argument. The heme iron of cytochrome c experiences a ligand exchange (replacing a methionine by, perhaps, a lysine sidechain) upon removal of water by lyophilization (*75*). Both, desiccation and freezing of catalase have similar consequences on the activity of the enzyme (*76*). Freezing and thawing appear to alter certain physical-chemical properties of some proteins and similar but greater changes are observed during freeze-drying, which is presumably a more drastic method for affecting the hydration shell (*77*). Thermal transitions and irreversible structural changes occur in the lipoprotein of egg yolk during freezing and these changes are strongly dependent on water content where water is reduced below a critical value of 20% (which corresponds approximately to the two "nonfreezable" classes of water in Table III) (*78*). There are, of course, numerous—mostly casual—references in the literature to protein denaturation, or enzyme inactivation, during frozen storage. These observations are frequently coupled with the suggestion that the extensive (or complete?) transition of water to ice affected or destroyed the hydration shell of the protein thus causing a structural change. These suggestions are for the most part based on inadequate experimental evidence.

"Unexplained" Effects. Some observations are difficult to attribute to any of the freezing mechanisms considered so far. Most amenable to conceptual accommodation are those cases where a particular reactive intermediate, or a particular conformer, appears to have been "trapped." This is believed to occur at very low temperatures where mobility at the atomic level is severely limited partly because of little thermal motion and partly because of the physical constraint provided by the glassy or icy structures of the medium. Normally unstable free radical derivatives of proteins can be studied by trapping them at low temperatures (e.g. *79*). This technique also can be used to stabilize what appear to be specific conformational isomers of normal structures. For example, this approach has enabled the detection of an ammonia–catalase complex (*80*), conformers of the iron proteins conalbumin and transferrin (*81*), and a conformer of the flavoprotein L-amino acid oxidase (*82, 83*).

An interesting special case is provided by aldolase. If frozen in the presence of alkylating agents, it undergoes inactivation by sulfhydryl alkylation (*84*). However, this inactivation appears to occur during thawing, conditioned supposedly by conformational stresses that develop during frozen storage.

Protein Alterations and Behavior Associated with Dynamic Aspects of Freezing

It seems appropriate to deal somewhat more explicitly with the dynamic aspects of the freezing process in contrast to the effects of a given state of a cold or frozen system. This is probably also the best place to acknowledge a major omission throughout this overview. None of the discussion has dealt specifically with the process of thawing. It must be taken for granted that some "freezing" effects are in reality "thawing" effects although in very few cases is there conclusive evidence on this point. It can be assumed that whatever arguments may be reasonably advanced regarding the dynamics of freezing and its consequences also may be advanced, by suitable extension, inversion, or inference, to the dynamics of the melting process. Undoubtedly, "melting effects" are real, however neglected their separate consideration may be.

Protein Antifreezes. An interesting group of proteins merits special notice not because freezing affects them but because they affect the freezing process. Fishes inhabiting polar waters have recently been shown to possess unusually structured, alanine-rich serum glycoproteins that depress the freezing point of aqueous systems by some noncolligative process (*85, 86*). Similar antifreeze proteins also have been found in the mussel (*87*). These proteins have a marked effect on the mode of ice growth, indicating that they may exert their freezing inhibition by becoming adsorbed onto particular facets of the growing ice crystal. The presence of antifreeze proteins in fish blood has been shown to vary according to an annual cycle (*88*). Circular dichroic measurements indicate that some antifreeze proteins are disordered in solution while others have a great helix content—suggesting that there is no particular conformation to which the antifreeze property can be attributed (*89*). Studies involving Raman spectra tend to confirm this suggestion and indicate also that these proteins do not affect the bulk properties of water or ice that exist in their presence (*90*). A synthetic alanine–aspartic acid copolymer—modeling the natural antifreeze proteins—was shown to depress the freezing point of water to about one-third the extent of the antifreeze proteins of fish (*91*). It is noteworthy that this polypeptide contained no carbohydrate sidechains as do the natural antifreeze proteins.

It is not known whether such natural cryoprotectants occur widely distributed among animal species. However, a relevant experiment was conducted with brain slices, from warm-adapted and hibernating hamsters, before and after freezing (*92*). Tissue slices from the hibernating hamster exhibited higher than normal oxygen consumption rates after freezing. This result did not occur with slices from the warm-adapted animal. These experiments prompted the suggestion that, in the hibernating animal, membrane proteins of the brain may be modified in a way that enhances their antifreeze potential.

Two early observations may have some relevance to the mechanism by which antifreeze proteins act. Polymer gels of highly ramified, net-like structures have been shown to depress the freezing point of water and cause the ice crystals to assume a dominant growth orientation (*93*). It is also noteworthy, perhaps, that certain amino acids are good ice nucleators whereas others are not (*94*).

Events Occurring at or Near the Water–Ice Interface. It would take us too far afield to attempt to deal in detail with theories and speculations concerning the small zone existing between supercooled liquid and the growing ice mass. But, in view of the potential importance this zone may have with regard to freezing effects, some discussion of it is appropriate. I already acknowledged the significance of this interface when consideration was given to the increase in solute concentration that occurs when solutes are rejected by the growing ice crystals. Also, the earlier discussion of factors that influence the formation of transiently existing, concentrated "puddles" had clear implications regarding this interface. I shall now consider this interface in more detail.

Of historical interest is a review of a century's worth of experiment and thought directed at the nature of the ice surface (*95*). A major conclusion in this review is that a transition layer of water molecules does in fact exist at the ice surface. This review is concerned, however, with interfaces between ice and the vapor phase or another solid phase. More directly pertinent to our purposes is another fascinating review that takes a detailed look at the interface "as seen from the liquid side" (*96*). This assessment of relevant evidence also results in the firm conclusion that an interface layer with special properties exists. Of particular importance to the present discussion is the notion that the transition layer consists of water molecules in an oriented, polar arrangement. The polarization of this layer is ascribed to the need to relieve strains in the ice lattice. These strains arise from the enlarging of the normal hydrogen–oxygen angles in order to fit the requirements of the tetrahedral structure in ice. This strain relief process would result, among other things, in the maintenance of an electric potential between the liquid and solid phases and would account for the experimentally observed selective incorporation of ions in the ice structure.

The notion of a polar, oriented interface layer of water molecules—while based on an ad hoc argument—is consistent with experimental data. It is in accord, for example, with the large electric potentials that develop across the water–ice interface during freezing of dilute salt solutions (*97*). This occurrence probably results from the often demonstrated selective incorporation of ions (in most cases, anions) into the ice (e.g. *98–106*).

Some time ago, I observed a freezing-induced conformational change in a protein that appeared to be associated with selective rejection of ions at the liquid–ice interface (*107*). The results of this study appeared to make a good case for the claim that such solute redistributions may have important effects on proteins during freezing, particularly since a very high concentration of protons may develop in the interface region. Although this concentration of protons is transient, it presumably persists for the duration of crystal growth.

The reality of an interface region in which solute concentrations may be high during the freezing process was shown by direct measurements. It was found that the interface layer may have a thickness on the order of 100–1000 nm, the exact value varying inversely with the rate of ice propagation (*108*).

It has also been suggested that as the ice surface grows, the solute-enriched layer just ahead of it will assume a lower freezing point than the solution of lower concentration that exists further away from the ice surface. This situation would be most likely to occur during rapid freezing when the diffusive redistribution of solute (rejected by the ice) is slow compared with the rate of its concentration build-up. This could lead to nucleation in advance of the freezing boundary, thereby causing entrapment of concentrated solution pools within the ice structure (*109*). It should be noted that on a gross scale this would encourage more uniform solute distribution in the ice but on a microscale (which would be relevant for molecular interactions) the system would be decidedly heterogeneous and concentration effects should be great.

It seems useful to conclude this section with data illustrating the effects of various solutes on the propagation rates of ice (Table IV). Discussion of the effects that solutes have on the relationship between ice growth rate and the degree of supercooling is available elsewhere (*111*). A more recent discussion of equilibrium and nonequilibrium aspects of freezing provides a most helpful aid in the integrated consideration of aqueous solutions subjected to freezing, with particular reference to biological systems (*112*).

Possible "Interface Effects" on Proteins. Interface effects that arise during growth of ice crystals can influence proteins but this issue is rarely addressed in a direct fashion. During freezing, the egg yolk protein phosvitin has been shown to undergo a major conformational change toward an ordered structure. This is caused by the transiently high concentration of protons shown to exist in the ice–liquid interface region

Table IV. Retardation of Ice Propagation by Diverse Solutes[a]

Solute	*Rate of Ice Growth (cm/sec) at Given Solute Concentration*[b] *0%*	*10%*
Acetates		
lithium	1.41	0.13
sodium	2.76	0.33
potassium	2.69	0.62
ammonium	1.74	0.35
Chlorides		
lithium	1.32	0.10
sodium	3.09	0.48
potassium	2.82	1.08
ammonium	2.09	0.54
Thiocyanates		
sodium	3.02	0.85
potassium	2.82	—
ammonium	2.09	0.96
Acids		
acetic	2.14	0.76
glycolic	2.14	0.83
malic	1.66	0.78
citric	1.29	0.60
aminoacetic	2.19	—
Alcohols		
methanol	1.20	0.25
ethanol	0.43	0.09
propanol	0.54	0.12
glycerol	1.38	0.41
Sugars		
glucose	1.02	0.38
fructose	1.20	0.46
sucrose	0.85	0.36
Proteins		
lysozyme	0.81	0.56
albumin	0.76	—
None (pure water)	ca. 7	—

[a] Original data from Lusena (*110*). In the original article, values were in logarithmic form. Above data were recalculated to give dimensions of velocity. All experiments carried out with solutions supercooled 10° below their respective freezing points. The data have *relative* significance.

[b] All experiments yielded a biphasic relationship between the logarithm of the rate of ice growth and the concentration of the solute. At low concentrations of all solutes, the rate of ice growth fell drastically within a narrow concentration range. Further rate diminution occurred over a range of higher concentrations where the log rate vs. concentration relationship was linear. Extrapolation of the linear portion of this linear relationship to zero concentration produces the values listed under the heading "0%." The data in the "10%" concentration column reflect growth rates which tend to be associated with the lower end of the concentration range in which the linear relationship holds.

during crystal formation (*113*). It is also noteworthy that a highly ordered polymer forms during freezing of a solution of polyadenylic acid. This transconformation may also be an "interface effect" (*114*).

When protein alterations occur as a result of experimental conditions that are expected to influence the dynamics of the freezing process, it seems reasonable to postulate that interface effects may be operative. However, such experiments are rarely designed for the purpose of elucidating the mechanism by which freezing causes protein alteration. Therefore, the mechanism remains largely a matter of conjecture. From among the many candidates for an interface effect, a few will be cited—essentially randomly—simply to underpin the suggestion that these effects may have a broader significance in the context of protein cryochemistry than is generally acknowledged.

The possibility of interface effects has been specifically noted in a study involving freezing of lipoamide dehydrogenase (*115*). This possibility was also mentioned in a study of conalbumin and transferrin, referred to earlier (*81*). An interface mechanism also may be operative with regard to freezing-induced alterations of phycoerythrin (*116*), lactic dehydrogenase (*117*), catalase (*118*), and other proteins, some of which were referred to earlier (*22*). Finally, it may be noted that interface effects, especially the transient distribution of solutes, may underlie some of the anomalous kinetics that have been described for certain reactions in the "frozen state" (*119*).

Conclusions

I set out to give an overview of freezing-induced alterations of proteins. I approached the task with a bias that, I hope, became clear as the discussion progressed. It seems to me that inquiries into protein behavior at low temperatures and in freezing systems are ripe for more sharply focussed studies of the mechanisms by which "freezing effects" operate. A massive amount of data exists concerning the consequences of low temperature exposure. A common deficiency is the failure to provide insight into underlying causes. Intellectually satisfying and practically useful generalizations are needed and these should emerge in the future. Indeed, the broad range of topics included in this symposium makes it evident that on-going investigations are effectively moving toward this goal.

Literature Cited

1. Meryman, H. T., Ed. "Cryobiology"; Academic: London and New York, 1966; pp vii–x.
2. Rey, L. R. *Ann. N. Y. Acad. Sci.* **1960,** *85,* 510.
3. Polge, C.; Smith, A. U.; Parkes, A. S. *Nature (London)* **1949,** *164,* 666.

4. Nash, T. In "Cryobiology"; Meryman, H. T., Ed.; Academic: London and New York, 1966; p 179.
5. Doucou, P. *Biochimie* **1971**, *53*, 1135.
6. Fink, A. L. *Acc. Chem. Res.* **1977**, *10*, 233.
7. Edelhoch, H.; Osborne, J. C., Jr. *Adv. Protein Chem.* **1976**, *30*, 183.
8. Brandts, J. F. *J. Am. Chem. Soc.* **1964**, *86*, 4302.
9. Brandts, J. F.; Hunt, L. *J. Am. Chem. Soc.* **1967**, *89*, 4826.
10. Brandts, J. F.; Oliveira, R. J.; Westort, C. *Biochemistry* **1970**, *9*, 1038.
11. Zipp, A.; Kauzmann, W. *Biochemistry* **1973**, *12*, 4217.
12. Franks, F. *Philos. Trans. R. Soc. London, Ser. B* **1977**, *278*, 89.
13. Franks, F.; Eagland, D. *CRC Crit. Rev. Biochem.* **1975**, *3*, 165.
14. Scheraga, H. A. *Ann. N. Y. Acad. Sci.* **1977**, *303*, 2.
15. Richards, F. M. *Annu. Rev. Biophys. Bioeng.* **1977**, *6*, 151.
16. Sizer, I. W.; Josephson, E. S. *Food Res.* **1942**, *7*, 201.
17. Kavanau, J. L. *J. Gen. Physiol.* **1950**, *34*, 193.
18. Maier, V. P.; Tappel, A. L.; Volman, D. H. *J. Am. Chem. Soc.* **1955**, *77*, 1278.
19. Hultin, E. *Acta Chem. Scand.* **1955**, *9*, 1700.
20. Tappel, A. L. In "Cryobiology"; Meryman, H. T., Ed.; Academic: London and New York, 1966; p 163.
21. Markert, C. L. *Science* **1963**, *140*, 1329.
22. Chilson, O. P.; Costello, L. A.; Kaplan, N. O. *Fed Proc., Fed. Am. Soc. Exp. Biol.* **1965**, *24*, S-55.
23. El-Negoumy, A. M. *J. Dairy Sci.* **1974**, *57*, 1170.
24. Hood, L. F.; Seifried, A. S. *J. Food Sci.* **1974**, *39*, 121.
25. Hasiak, R. J.; Vadehra, D. V.; Baker, R. C.; Hood, L. *J. Food Sci.* **1972**, *37*, 913.
26. Smith, M. B.; Back, J. F. *Biochim. Biophys. Acta* **1975**, *388*, 203.
27. Sato, Y.; Aoki, T. *Agric. Biol. Chem.* **1975**, *39*, 29.
28. Sato, Y.; Takagaki, Y. *Agric. Biol. Chem.* **1976**, *40*, 49.
29. Chang, C. H.; Powrie, W. D.; Fennema, O. *J. Food Sci.* **1977**, *42*, 1658.
30. Parkinson, T. L. *J. Sci. Food Agric.* **1977**, *28*, 806.
31. Grek, A. M.; Lugovoi, V. I.; Bondarenko, T. P.; Morozov, I. A. *Aktual. Vopr. Kriobiol. Kriomed., Mater. Simp.* **1974**, 43; *Chem. Abstr.* **1976**, *84*, 146529.
32. Skrabut, E. M.; Crowley, J. P.; Catsimpoolas, N.; Valeri, C. R. *Cryobiology* **1976**, *13*, 395.
33. Jackson, W. M.; Kostyla, J.; Nordin, J. H.; Brandts, J. F. *Biochemistry* **1973**, *12*, 3662.
34. Araki, T. *Biochim. Biophys. Acta* **1977**, *496*, 532.
35. Dzhafarov, A. I.; Kol's, O. R. *Biofizika* **1976**, *21*, 653; *Chem. Abstr.* **1976**, *85*, 107064.
36. Stahl, W. L.; Swanson, P. D. *Neurobiology* **1975**, *5*, 393.
37. Van den Berg, L.; Rose, D. *Arch. Biochem. Biophys.* **1959**, *81*, 319.
38. Van den Berg, L. *J. Dairy Sci.* **1961**, *44*, 26.
39. Van den Berg, L. *Cryobiology* **1966**, *3*, 236.
40. Soliman, F. S.; Van den Berg, L. *Cryobiology* **1971**, *8*, 73.
41. Van den Berg, L.; Soliman, F. S. *Cryobiology* *1969*, *6*, 93.
42. Pincock, R. E.; Lin, W.-S. *J. Agric. Food Chem.* **1973**, *21*, 2.
43. Sheppard, H.; Tsien, W. H. *Biochim. Biophys. Acta* **1974**, *341*, 489.
44. Janson, J. C.; Ersson, B.; Porath, J. *Biotechnol. Bioeng.* **1974**, *16*, 21.
45. Orii, Y.; Morita, M. *J. Biochem. (Tokyo)* **1977**, *81*, 163.
46. Posey, J. C.; Smith, H. A. *J. Am. Chem. Soc.* **1957**, *79*, 555.
47. Eigen, M.; DeMaeyer, L. *Proc. R. Soc. London, Ser. A* **1958**, *247*, 505.
48. VonHippel, A. R.; Runck, A. H.; Westphal, W. B. *Gov. Rep. Announce. (U.S.)* **1974**, *74*, 59.
49. Szent-Gyorgyi, A. *Science* **1956**, *124*, 873.

50. Pincock, R. E. *Ace. Chem. Res.* **1969,** *2,* 97.
51. Wang, S. Y. *Nature (London)* **1961,** *190,* 690.
52. Horne, R. A. *J. Inorg. Nucl. Chem.* **1963,** *25,* 1139.
53. Luyet, B. J. *Biodynamica* **1957,** *7,* 293.
54. Macklin, W. C.; Ryan, B. F. *J. R. Meteorol. Soc.* **1962,** *88,* 548.
55. Orii, Y.; Iizuka, T. *J. Biochem. (Tokyo)* **1975,** *77,* 1123.
56. Levitt, J. In "Cryobiology"; Meryman, H. T., Ed.; Academic: London and New York, 1966; p 495.
57. Levitt, J. *Cryobiology* **1966,** *3,* 243.
58. Khan, A. W.; Davidkova, E.; Van den Berg, L. *Cryobiology* **1968,** *4,* 184.
59. Ishiwata, S. *J. Biochem. (Tokyo)* **1976,** *80,* 595.
60. Cherry, J. P.; Ory, R. L. *J. Agric. Food Chem.* **1973,** *21,* 656.
61. Weber, G. *Adv. Protein Chem.* **1975,** *29,* 1.
62. Fink, A. L. *Biochemistry* **1976,** *15,* 1580.
63. Makinen, M. W.; Fink, A. L. *Annu. Rev. Biophys. Bioeng.* **1977,** *6,* 301.
64. Petsko, G. A. *J. Mol. Biol.* **1975,** *96,* 381.
65. Franks, F. *Philos. Trans. R. Soc. London, Ser. B* **1977,** *278,* 33.
66. Timasheff, S. N. *Acc. Chem. Res.* **1970,** *3,* 62.
67. Kuntz, I. D., Jr.; Kauzmann, W. *Adv. Protein Chem.* **1974,** *28,* 239.
68. Shlafer, M. *Fed. Proc., Fed. Am. Soc. Exp. Biol.* **1977,** *36,* 2590.
69. Hagler, A. T.; Scheraga, H. A.; Nemethy, G. *Ann. N. Y. Acad. Sci.* **1973,** *204,* 51.
70. Eisenberg, D.; Kauzmann, W. "The Structure and Properties of Water"; Oxford University Press: New York and Oxford, 1969; pp 254 ff.
71. Drost-Hansen, W. *Ann. N. Y. Acad. Sci.* **1973,** *204,* 100.
72. Sussman, M. V.; Chin, L. *Science* **1966,** *151,* 324.
73. Parungo, F. P.; Wood, J. *J. Atmos. Sci.* **1968,** *25,* 154.
74. Austin, R. H.; Beeson, K. W.; Eisenstein, L.; Frauenfelder, H.; Gunsalus, I. C. *Biochemistry* **1975,** *14,* 5355.
75. Aviram, I.; Schejter, A. *Biopolymers* **1972,** *11,* 2141.
76. Darbyshire, B. *Cryobiology* **1974,** *11,* 148.
77. Hanafusa, N. *Bull. Inst. Int. Froid, Annexe* **1973,** *5,* 9; *Chem. Abstr.* **1974,** *81,* 165554.
78. Smith, M. B.; Back, J. F. *Biochim. Biophys. Acta* **1975,** *388,* 203.
79. Bray, R. C.; Lowe, D. J.; Capeillere-Blandin, C.; Fielden, E. M. *Biochem. Soc. Trans.* **1973,** *1,* 1067.
80. Ehrenberg, A.; Estabrook, R. W. *Acta Chem. Scand.* **1966,** *20,* 1667.
81. Price, E. M.; Gibson, J. F. *J. Biol. Chem.* **1972,** *247,* 8031.
82. Curti, B.; Massey, V.; Zmudka, M. *J. Biol. Chem.* **1968,** *243,* 2306.
83. Coles, C. J.; Edmondson, D. E.; Singer, T. P. *J. Biol. Chem.* **1977,** *252,* 8035.
84. Friedrich, P.; Foldi, J.; Varadi, K. *Acta Biochim. Biophys.* **1974,** *9,* 1.
85. Raymond, J. A.; Lin, Y.; DeVries, A. L. *J. Exp. Zool.* **1975,** *193,* 125.
86. Raymond, J. A.; DeVries, A. L. *Proc. Natl. Acad. Sci. U.S.A.* **1977,** *74,* 2589.
87. Theede, H.; Schneppenhelm, R.; Beress, L. *Mar. Biol.* **1976,** *36,* 183.
88. Fletcher, G. L. *Can. J. Zool.* **1977,** *55,* 789.
89. Raymond, J. A.; Radding, W.; DeVries, A. L. *Biopolymers* **1977,** *16,* 2575.
90. Tomimatsu, Y.; Scherer, J. R.; Yeh, Y.; Feeney, R. E. *J. Biol. Chem.* **1976,** *251,* 229.
91. Ananthanarayanan, V. S.; Hew, C. L. *Nature (London)* **1977,** *268,* 560.
92. Tower, D. B.; Goldman, S. S.; Young, O. M. *J. Neurochem.* **1976,** *27,* 285.
93. Kuhn, W.; Bloch, R.; Laeuger, P. *Kolloid-Z.* **1963,** *193,* 1.
94. Barthakur, N.; Maybank, J. *Nature (London)* **1963,** *200,* 866.
95. Jellinek, H. H. G. *J. Colloid Interface Sci.* **1967,** *25,* 192.
96. Drost-Hansen, W. *J. Colloid Interface Sci.* **1967,** *25,* 131.

97. Workman, E. J.; Reynolds, S. E. *Phys. Rev.* **1950**, *78*, 254.
98. Lodge, J. P.; Baker, M. L.; Pierrard, J. M. *J. Chem. Phys.* **1956**, *24*, 716.
99. Jaccard, C.; Levi, L. *Z. Angew. Math. Phys.* **1961**, *7*, 70.
100. Heinmets, F. *Trans. Faraday Soc.* **1962**, *58*, 788.
101. DeMicheli, S. M.; Iribarne, J. V. *J. Chim. Phys., Phys.-Chim. Biol.* **1963**, *60*, 767.
102. Gross, G. W. *J. Geophys. Res.* **1965**, *70*, 2291.
103. Parreira, H. C.; Eydt, A. J. *Nature (London)* **1965**, *208*, 33.
104. Anantha, N. G.; Chalmers, B. *J. Appl. Phys.* **1967**, *38*, 4416.
105. LeFebre, V. *J. Colloid Interface Sci.* **1967**, *25*, 263.
106. Pruppacher, H. R.; Steinberger, E. H.; Wang, T. L. *J. Geophys. Res.* **1968**, *73*, 571.
107. Taborsky, G. *J. Biol. Chem.* **1970**, *245*, 1063.
108. Terwilliger, J. P.; Dizio, S. F. *Chem. Eng. Sci.* **1970**, 1331.
109. Himes, R. C.; Miller, S. E.; Mink, W. H.; Goering, H. L. *Ind. Eng. Chem.* **1959**, *51*, 1345.
110. Lusena, C. V. *Arch. Biochem. Biophys.* **1955**, *51*, 277.
111. Pruppacher, H. R. *J. Colloid Interface Sci.* **1967**, *25*, 285.
112. MacKenzie, A. P. *Philos. Trans. R. Soc. London, Ser. B* **1977**, *278*, 167.
113. Taborsky, G. *J. Biol. Chem.* **1970**, *245*, 1054.
114. Zimmerman, S. B.; Coleman, N. F. *Biopolymers* **1972**, *11*, 1943.
115. Visser, J. *Meded. Landbouwhogesch. Wageningen* **1970**, *70–7*, 90.
116. Leibo, S. P.; Jones, R. F. *Arch. Biochem. Biophys.* **1964**, *106*, 78.
117. Greiff, D.; Kelly, R. T. *Cryobiology* **1966**, *2*, 335.
118. Fishbein, W. N.; Winkert, J. W. *Cryobiology* **1977**, *14*, 389.
119. Oakenfull, D. G. *Aust. J. Chem.* **1972**, *25*, 769.

RECEIVED June 16, 1978.

2

Properties of Protein–Water Systems at Subzero Temperatures

I. D. KUNTZ

Department of Pharmaceutical Chemistry, University of California, San Francisco, CA 94143

The hydration of proteins at subzero temperatures is reviewed. The thermodynamics of the protein–water system and the water molecule dynamics are discussed. The hydration layer around a protein at low temperature is best thought of as being in a glass-like state with the water molecules selectively oriented near ionic and polar groups at the protein surface. Water motions in the nanosecond and microsecond range have been detected.

This paper will summarize the available information on a number of properties of protein–water systems below 0°C. We will direct our attention to thermodynamic data and to experiments that speak to the dynamics of the systems. In some cases, we use results at higher temperatures to infer low-temperature behavior. Our primary purpose is to organize the experimental results. No single model is likely to cover all the facts.

Thermodynamic Properties

Phase Diagram. We will treat protein–water systems as two-component systems although many preparations contain buffer or supporting electrolytes that formally require three (or more) components. These additional components can always be expected to influence a number of measurements and will alter the entire phase diagram, including the bulk melting behavior, in complex ways. We will call attention to such effects where appropriate.

0-8412-0484-5/79/33-180-027$5.00/0

The first important observation is that any protein–water system will contain some water molecules that do not freeze into a crystalline state at temperatures far below zero (*1, 2, 3*). Although the lowest temperatures reached vary from one technique to another (ca. $-70°C$ for NMR and $-196°C$ for ir) no evidence for a first-order crystallization for the "nonfreezing" water has been reported. Further, annealing procedures do not induce crystallization, suggesting strongly that a state with some degree of stability has been achieved. For most proteinaceous systems, the starting preparation contains water in excess of the nonfreezing water, and the excess water readily freezes to a normal ice phase if no low molecular weight additives are present. For preparations with bulk water contents less than 0.3–0.5 g water/g protein, no ice formation is detectable by NMR at any low temperature. The development of very small subnucleation crystalloids of ice has been suggested in some cases (*5, 6*).

Water–protein systems show distinctly different freezing behavior than aqueous solutions that contain many low molecular weight components. A dilute NaCl solution, for example, will have liquid water in stable equilibrium with ice. The amount of unfrozen water will be quite temperature-dependent, falling smoothly until the eutectic point is attained, which for NaCl is $-21°C$. At a degree or so below this temperature, no water signal can be detected by conventional NMR techniques, and ir will indicate a mixture of ice and $NaCl \cdot H_2O$, a well characterized crystalline monohydrate. Dilute aqueous protein solutions will also show ice formation beginning at about zero degrees, with the amount of unfrozen water decreasing rapidly with decreasing temperature until a value in the 0.3–0.5 g H_2O/g protein range is reached (usually at about $-10°C$ in the absence of salts). However, in contrast to the eutectic behavior described above for NaCl, the amount of frozen water then remains essentially constant with further decreases in temperature. Protein crystals—normally hydrated at roughly 1 g water/g protein—show similar behavior to aqueous solutions containing the same protein composition. (However, some exceptions to this pattern have been reported (*7, 8, 9*). These include tropomyosin (a muscle protein) and a number of polypeptide solutions that show a continual decrease in the fraction of unfrozen water, at least down to the low temperature limit of NMR.) No simple explanation has been offered for these two types of behavior. Perhaps the most straightforward qualitative explanation is that conventional eutectic behavior arises because of the intersection of two phase boundaries: the freezing point curve that represents equilibrium conditions between the solution and ice and the melting point curve that represents equilibrium conditions between the solution and (generally) a well defined crystalline hydrate. Polymer solutions have well characterized freezing point curves, but they usually do not form well behaved crystalline hydrates. Thus the lack of a eutectic point is not too surprising. The slope of the freezing point curve at temperatures far below zero is not easily determined but it is thought to depend on a number of parameters such as ΔH of fusion and the Flory-Huggins interaction constant, and these could account for the range of behavior described above. We know that certain solutions containing low molecular weight hydroxylic solutes (sugars, alcohols, etc.) can form glasses at low temperatures. Protein solutions at high polymer concentrations and/or low temperatures have many features in common with these glasses.

For systems in which the unfrozen water content is independent of temperature, the amount of unfrozen water at any given subfreezing temperature is related to the ionic composition of the macromolecule—the more ionic, the more unfrozen water observed (*4, 8*). Negative ions are found to be hydrated more extensively than positive ions (*8*). Preparations containing oppositely charged macromolecules (e.g., ribosomes, viruses, and membrane proteins) frequently are less hydrated than would be predicted based on the behavior of the fixed components (*4*). This suggests that ionic groups become buried in a mixed system. I am not aware of a general treatment of counter-ion effects on macromolecular hydration. The widely varying eutectic points in such systems make a definitive study quite difficult. A phase diagram for water–protein systems has been proposed (*4*) to summarize the above discussion.

Enthalpy, Entropy, and Heat Capacity of Protein–Water Systems Below 0°C. A number of investigators have reported the apparent enthalpy of fusion as a function of temperature and composition for several hydrated proteins. MacKenzie and coworkers (*10*) determined absorption isotherms at low temperatures and found that: 1) these absorption isotherms have essentially the same sigmoidal shapes as those observed above zero degrees; 2) the magnitudes of the values for partial molal enthalpy and entropy increase as the content of unfrozen water decreases; 3) the heat of fusion decreases as the content of unfrozen water decreases; and 4) the heat capacity of the system increases as the content of unfrozen water increases. Taking these findings all together, the thermodynamic properties of "unfrozen" water are not very different from those of supercooled water at comparable temperatures.

One should always remember that it is difficult to proceed from thermodynamic measurements to molecular properties. This is particularly true in multicomponent systems. That is, water–water and water–protein interactions are inextricably mixed with protein–protein interactions when one measures partial molal quantities. Assumptions and experiments beyond those yielding thermodynamic information are needed to determine what is happening at the molecular level. The major cause for concern, here, is the real possibility that significant changes in protein conformation can occur at low temperatures. There is some evidence, for example, that protein denaturation is important in some cases (*11*). Clearly, the concentrations of any solutes increase dramatically as bulk water freezes. To the extent that appreciable denaturation or structural modification occurs, one can expect difficulties in interpreting thermodynamic measurements on these systems.

In summary, I suggest that the phenomenon of nonfreezing water in protein–water systems at low temperature might arise from the lack of equilibrium with a well defined crystalline protein hydrate phase. For proteins in which a crystalline phase is known to exist, MacKenzie has shown that samples can be taken through the crystallization region with no crystal formation (*10*). This lack of equilibrium is quite likely a kinetic problem because of the very high viscosity present. Surface effects might also contribute significantly since the surface area per unit volume is very large for any molecularly dispersed polymer. On the whole, the thermodynamic properties reported for nonfreezing water appear rather similar to those of supercooled water.

Motions in Protein–Water Systems

The most powerful technique for studying molecular motions in protein–water systems below 0°C is magnetic resonance. Dielectric relaxation measurements can be used, but these measurements are more suitable at higher temperatures in homogenous solutions (*13*). Recently, the frequency dependence of the mehcanical properties of biopolymers has been shown to yield considerable kinetic information (*14*). I will limit discussion to the salient results attainable from these techniques.

NMR. Magnetic resonance experiments at low temperatures have been limited largely to proton and deuteron NMR of the water molecules in water–polymer preparations. This is reasonable because of the sensitivity attainable and because the most rapidly moving molecular species (water) is the most easily detected. I will discuss only systems without macroscopic order (e.g., frozen solutions of globular proteins) but the interested reader will find intriguing reports of nmr measurements on protein crystals and on fibrous or layered materials (*15, 16, 17*).

Important results that have been obtained from NMR analysis of protein solutions at subzero temperatures are:

(1) The water proton linewidth at -20°C is 100 times sharper for water than for ice, and even at -60°C the difference is very marked.

(2) All systems we have studied to date show a minimum spin-lattice relaxation time (T_1) or a maximum spin-lattice relaxation rate ($1/T_1$) at ca. -35°C for a larmor frequency of 40 MHz. The minimum for T_1 moves to lower temperatures as the larmor frequency decreases.

(3) In our work, the proton linewidth is approximately equal to the spin-spin relaxation rate ($1/T_2$) and this relaxation is noticeably faster than $1/T_1$ at all subzero temperatures.

(4) The spin-lattice relaxation rate ($1/T_1$) is reported to be quite frequency-dependent (see (2) above), and rotating frame experiments ($1/T_{1\rho}$) at low temperatures also indicate frequency dependent relaxations (*18, 19*).

(5) No rapid motion in the macromolecular components has been reported.

Certain conclusions can be drawn from these observations although detailed interpretation will depend on future quantitative developments. First, most of the water molecules detected by NMR are moving quite rapidly at subzero temperatures. The sharp NMR signal for protons of unfrozen water is perhaps the best evidence that a noncrystalline state of water is present in these samples. Quantitative conclusions about proton mobility can be derived from the relaxation data given above. Without going into detail, above $-35°C$ the average rotational correlation time for water protons is certainly less that 4×10^{-9} seconds. In our early papers we expressed the view that this average correlation time applied to most of the water molecules that contribute to the nmr signal. It is now clear that there is no hard evidence in support of this position. The T_1 minimum is broad and not as deep as simple theory (*20*) would predict. Contributions from spin-spin diffusion are readily detected (*17, 21, 22*). There is a real chance that a very significant fraction of the unfrozen water molecules have rotational correlation times considerably shorter than 10^{-9} seconds. The correlation time for supercooled water at these temperatures is about 3×10^{-10} seconds. The definitive experiment would involve measurement of high frequency dielectric dispersion in these materials.

Second, observations (3) and (4) suggest thas some of the water molecules exhibit slow motions (e.g., slower than 4×10^{-9} seconds). There are two important restrictions here. It is not possible to be very quantitative in discussing slow motions because there is a substantial amplification factor for the way slow motions influence NMR linewidth. Also, there is no reason to assume that a given water molecule experiences a single rotational motion. It is much more likely that any given molecule samples all possible motions for varying periods of time. Observation of nonexponential decay of NMR signals is not, in itself, evidence of slowly exchanging populations of water molecules (*17, 21, 22*). Thus, the conservative conclusion is that a small fraction of the water motions at, say, $-35°C$ are considerably slower than 10^{-9} and might approach 10^{-6} seconds.

Third, there is no NMR evidence for rapid tumbling of the macromolecules or for rapid motion of their sidechains. However, these motions would be difficult to detect. Careful experiments of protein relaxation in concentrated solutions and gels are needed before accurate conclusions can be drawn with regard to motions of macromolecules.

Thus we interpret the NMR results to date as providing reasonable evidence for two classes of motion in frozen protein solutions in the subzero range. The first and most prominent motion is the quite rapid tumbling of nonfrozen water molecules. The second, more poorly defined motion involves many fewer water protons at any instant of time and these have correlation times of approximately 10^{-5} to 10^{-7} seconds in the temperature range of -20 to $-50°C$.

Dielectric Measurements. Recent dielectric experiments at above-zero temperatures have detected a number of molecular motions that might occur at low temperatures (*23, 24, 25*). In particular, sidechain motions and the motions of counter-ions could contribute to the water relaxation processes. Frequencies in the range of 10^4 to 10^8 have been suggested for such effects, but, at present, there is no direct line of evidence connecting these motions to the low frequency nmr results.

Mechanical Properties. Hiltner and coworkers have measured the dynamic properties of hydrated proteins and polypeptides at low temperatures (*6, 14*). This technique involves direct mechanical deformation and has a very low frequency "window" for motions at 1 Hz. Collagen, for example, shows two processes, the faster one moving through the 1 Hz window at approximately $-125°C$ while the slower process is seen at $-75°C$. Hiltner suggests that the first motion is related to polymer sidechain effects, while the second is assigned to a specific water–protein interaction. A large activation barrier would move these motions into the NMR frequency range at higher temperatures.

In summary, a substantial number of possible motions exist for water molecules associated with proteinaceous materials at low temperatures. A very wide range of frequencies exists: from a few Hz to a few GHz. A combination of studies, involving NMR measurements of the frequency dependence of relaxation rates and the dielectric and mechanical techniques described above, will be required to characterize and assign all these motions. Our present interpretation is that the water motions appear to reflect the total spectrum of kinetic events in the system.

Acknowledgment

Support from the National Institutes of Health and the Division of Research Resources is gratefully acknowledged.

Literature Cited

1. Migchelsen, C.; Brendsen, H. J. C.; Rupprecht, A. *J. Mol. Biol.* **1968,** *37,* 235.
2. Kuntz, I. D.; Brassfield, T. S.; Law, G. D.; Purcell, G. V. *Science* **1969,** *163,* 1329.
3. Falk, M.; Poole, A. G.; Goymour, C. G. *Can. J. Chem.* **1970,** *48,* 1536.
4. Kuntz, I. D.; Kauzmann, W. *Adv. Protein Chem.* **1974,** *28,* 239.
5. Luyet, B. J. *Ann. N. Y. Acad. Sci.* **1965,** *125,* 502.
6. Nomura, S.; Hiltner, A.; Lando, J. B.; Baer, E. *Biopolymers* **1977,** *16,* 231.
7. Blanshard, J. M. V.; Derbyshire, W. In "Water Relations of Foods"; Duckworth, R. B., Ed.; Academic: New York, 1975; p 559.
8. Kuntz, I. D. *J. Amer. Chem. Soc.* **1971,** *93,* 514.
9. Ramirez, J. E.; Cavanaugh, J. R.; Purcell, J. M. *J. Phys. Chem.* **1974,** *78,* 80.
10. MacKenzie, A. P. In "Water Relations in Foods"; Duckworth, R. B., Ed.; Academic: New York, 1975; p 477.
11. Kuntz, I. D.; Brassfield, T. S. *Arch. Biochem. Biophys.* **1971,** *142,* 660.
12. Resing, H. A.; Wage, C. G. *ACS Symp. Ser.* **1976,** *34.*
13. Takashima, S.; Fishman, H. H., Eds., "Electric Properties of Biological Polymers, Water, and Membranes", *Ann. N. Y. Acad. Sci.* **1977,** Vol. *303.*
14. Shiraisi, H.; Hiltner, A.; Baer, E. *Biopolymers* **1977,** *16,* 2801.
15. Resing, H. A.; Garroway, A. N.; Foster, K. R. *ACS Symp. Ser.* **1976,** *34. 516.*
16. Woessner, D. E.; Snowden, B. S. *J. Colloid Interface Sci.* **1970,** *34,* 290.
17. Bryant, R. G. *Adv. Biophys. Bioengr.* **1978,** Vol. 8.
18. Kuntz, I. D.; Zipp, A.; James, T. L. *ACS Symp. Ser.* **1976,** *34.* 499.
19. Zipp, A; Kuntz, I. D.; James, T. L. *Arch. Biochem. Biophys.* **1977,** *178,* 435.
20. Bloembergen, N.; Purcell, E. M.; Pound, R. V.; *Phys. Rev.* **1948,** *73,* 679.
21. Edzes, H. T.; Samulski, E. T. *Nature* **1977,** *265,* 521.
22. Kimmich, R.; Noack, F. *Ber. Bunsenges Phys. Chemie.* **1971,** *75,* 269.
23. Cole, R. H. *Ann. N. Y. Acad. Sci.* **1977,** *303,* 59.
24. Mandel, M. *Ann. N. Y. Acad. Sci.* **1977,** *303,* 74.
25. Minakata, A. *Ann. N. Y. Acad. Sci.* **1977,** *303,* 107.

RECEIVED June 16, 1978.

3

Enzyme-Catalyzed Reactions in Unfrozen, Noncellular Systems at Subzero Temperatures

ANTHONY L. FINK

Division of Natural Sciences, University of California, Santa Cruz, CA 95064

Methods used to determine the effects of both cryosolvents and low temperatures on the catalytic and structural properties of enzymes are detailed, along with representative results. Observed effects of subzero temperatures on the structure of enzymes include increased association of oligomers, minor temperature-induced structural changes, and in most cases no detectable effects. Examples demonstrating the very good correspondence between the kinetics observed at subzero temperatures and those for the corresponding reaction techniques, are given. Systems illustrating changes in the rate-determining step and ΔH *with a decrease in temperature, and the potential of cryoenzymology to provide details about individual intermediates and their interconversions during catalysis, are presented.*

Many biochemical processes involve very rapid reactions and transient intermediates. Frequently the rapidity of the reaction causes major technical difficulties in ascertaining the details of the events occurring in the process. One approach to overcome this inherent problem is to utilize the fact that most chemical reactions are temperature dependent. This relationship is quantitatively described by the Arrhenius equation, $k = Ae^{-E_a}/RT$, where k represents the rate constant, A is a constant (the frequency factor), and E_a is the energy of activation. Consequently, by initiating the reaction at a sufficiently low temperature, interconversion of the intermediates may be effectively stopped and they may be accumulated and stabilized individually. Although the focus of this article is on the application of this low-temperature approach to the study of enzyme catalysis, that is, cryoenzymology, the technique is potentially of much wider biological application (*1, 2, 3*).

0-8412-0484-5/79/33-180-035$5.00/0

Because of their catalytic function, which provides one with an additional "handle" or probe for detecting structural effects, enzymes are particularly well suited for studying the behavior of proteins at low temperatures. In this article the emphasis will be on illustrating the effect of subzero temperature on both the structural and catalytic properties of the enzymes and the ability to accumulate, stabilize, and characterize intermediates on the catalytic reaction pathway with very low temperatures. Because the low-temperature effects are intimately related to the cryosolvents used, a brief discussion of the effects of the organic cosolvents is included.

Although the examples used to illustrate the points of this article have been drawn mostly from relatively simple hydrolytic enzymes studied in the author's laboratory, over 25 different enzymes have currently been subjected to cryoenzymological investigations at several laboratories (*6*). Although the first reports of enzyme-catalyzed reactions at subzero temperatures in fluid aqueous organic solvents appeared 25 years ago, it is only since the pioneering studies of Douzou, beginning about 10 years ago, that more comprehensive investigations have been performed. Considerable impetus to the development of the approach was given by the elegant studies of Douzou and coworkers (*2*) on the horseradish peroxidase system, which demonstrated the feasibility of "temporally resolving" discrete enzyme-substrate intermediates and the accumulation and characterization of individual intermediates.

A comprehensive qualitative and quantitative understanding of enzyme catalysis has been a long-standing goal of biochemistry. A necessary requirement to achieve this is a detailed knowledge of all the intermediates and transition-state structures along the reaction pathway. The technique of cryoenzymology has the potential to provide much of this information. Because several recent reviews covering various aspects of the technique are available (*2, 4–9*), this article will emphasize some of the effects of subzero temperatures observed on the catalytic and structural properties of select enzymes.

Cryoenzymology utilizes the following features of enzyme catalysis: the existence on the catalytic reaction pathway of several enzyme–substrate (or product) intermediate species, typically separated by energy barriers with enthalpies of activation of 7 to 20 kcal mol^{-1}; and the fact that the energies (enthalpies) of activation for the individual steps in the overall catalytic pathway are usually significantly different. For such elementary steps temperatures of $-100°C$ will result in rate reductions on the order of 10^5 to 10^{11} compared to those at 25 or 37°C (*5*). The theoretical basis of cryoenzymology has been presented in detail elsewhere (*5, 7, 9, 10*). If the reaction is initiated by mixing enzyme and substrate at a suitably low temperature, only the initial noncovalent ES complex will be formed. This is a result of there being insufficient energy available to overcome the energy barrier to the subsequent intermediate. If the temperature is gradually increased, a point will be reached where ES is transformed into the following intermediate (I_1). Maintenance or reduction of the temperature will allow this intermediate to be trapped. Further raising of the temperature will result in transformation of I_1 to a subsequent intermediate (I_2), and so on until the overall rate-limiting step is reached, at which point turnover will occur. Any intermediate whose rate of formation is more rapid than its rate of breakdown may be accumulated in this manner.

The major advantages unique to cryoenzymology stem from the potential to accumulate essentially all of the enzyme in the form of a particular intermediate. The large rate reductions allow the most specific substrates to be used and hence provide the most accurate model for the in vivo catalyzed reactions. Virtually all the standard chemical and biophysical techniques used in studying proteins and enzymes under normal conditions may be used at subzero temperatures. The main limitations of the technique are the necessity to use aqueous organic cryosolvent systems to prevent the inherent rate-limiting enzyme-substrate diffusion of frozen solutions, and the possibility that the potential-energy surface for the reaction may be such that conditions in which an intermediate accumulates cannot be attained.

The general approach that has been developed in the author's laboratory for cryoenzymological studies involves the following: (1) Selection of a suitable cryosolvent and demonstration that it causes no adverse effects on either the catalytic or structural properties of the enzyme (e.g., *11*); (2) The detection of intermediates at subzero temperature by monitoring suitable spectral probes in the substrate or enzyme, and the kinetic and thermodynamic characterization of the intermediates (e.g., *12*); and (3) The acquisition of structurally related data concerning the trapped intermediates, by techniques such as X-ray diffraction and nmr (nuclear magnetic resonance) (e.g., *13*). Details concerning the experimental methodology for experiments using proteins at subzero temperatures may be found in the following references: *5, 6, 8, 9, 11, 12*.

Cryosolvents

At present the most versatile method of obtaining a fluid solution of protein at subzero temperatures seems to be that of using aqueous organic solvent mixtures. A number of such solvent systems are known with freezing points in the vicinity of $-100°C$. These usually contain 60–80% of the organic component. The most useful cryosolvents are

Table I. Physical-Chemical Properties of Cryosolvents[a]

Cryosolvent	Freezing Point (°C)	Dielectric Constant at 0°	−50°	−100°C	pH* at 20°C of pH 4.75 Acetate[b]	Viscosity (cps)
Dimethyl sulfoxide						
50%	s.c.[c]	84	105	132	5.40	
65%	s.c.[d]	79	98	124	6.7	
Methanol						
50%	−49	68	—	—	5.45	38 @ −40°
70%	−85	57	71	—	5.95	48 @ −60°
80%	−100	49	66	88	6.1	
Ethylene glycol–methanol						
40%:20%	−71	67	88	—	5.60	420 @ −60°
10%:60%	s.c.[c]	59	79		6.20	76 @ −60°
Dimethyl formamide						
80%	−100	56	70	86	6.9	
Ethylene glycol						
50%	−44	72	93	—	5.25	125 @ −40°

[a] Based on Ref. *29* and *30*.
[b] The apparent pH of the mixed aqueous organic solvent when the aqueous component was pH 4.75.
[c] s.c. = super cooled.
[d] Fluid to $< -90°C$.

those based on methanol, ethanol, dimethyl sulfoxide, dimethyl formamide, and ethylene glycol–methanol. Extensive physical-chemical studies of these solvents have been carried out by Douzou and coworkers (*14, 15*). In general the viscosity, the pH* (the apparent pH in the cryosolvent), and the dielectric constant increase with decreasing temperature. Representative data for some common cryosolvent systems are shown in Table I.

Cosolvent Effects

Structural. Experimentally, one of the most noticeable features caused by the presence of organic cosolvents on protein structure is the decrease in the temperature at which denaturation occurs. Interestingly, for most of the enzymes thus far studied, at the 60 to 80% cosolvent concentration required in the cryosolvents, the midpoint of the thermal denaturation transition is usually in the $-10°$ to $+10°C$ range in the pH* region of catalytic activity. This means that in such solutions the enzymes are usually denatured at room temperature, but are in their native states at subzero temperatures. The effect of increasing cosolvent concentration on the midpoint of the thermal denaturation of ribonuclease A is given in Table II. As the concentration of ethanol increases the T_m moves to increasingly lower temperatures. This behavior seems quite typical (*3*).

In general the effect of cosolvent on the structure of a protein may be determined by examining the intrinsic uv spectral properties of the protein as the cosolvent concentration increases. Smooth monotonic or linear curves reflect solvent effects on the exposed aromatic residues, whereas sharp breaks in such plots occur if a structural perturbation occurs (Figure 1). For example, if the effect of increasing dimethyl sulfoxide concentration on β-galactosidase is examined at 0°C, pH* 7.0, by either fluorescence or absorption spectroscopy, smooth curves are observed up to and including 50% dimethyl sulfoxide, whereas a sharp break in these curves becomes apparent at higher concentrations (Figure 1) (*18*). The intrinsic fluorescence, absorbance, and circular dichroism spectra are most convenient for such studies (*11, 16, 17, 18*). Additional techniques that can be used for this purpose include proton nmr [e.g., the nmr spectrum for subtilisin in 65% dimethyl sulfoxide, and for ribonuclease A in 50 or 70% methanol, is essentially the same as in aqueous solution under otherwise similar conditions (Fink and Kar, unpublished observations)] and X-ray crystallography, *vide infra* (*13*). Based on the accumulated data for some 15 enzymes examined in the author's laboratory, we are able to state that for each enzyme investigated there is considerable evidence to indicate that for suitable cryosolvents the cosolvent has little detectable effect on the structure at appropriately low temperatures and definitely causes no major changes in the conformation of the protein. These statements apply to oligomeric as well as monomeric enzymes, for example, β-galactosidase, β-glucosidase, and

Table II. Effect of Cosolvents on the Midpoint (T_m) of the Reversible Native ⇄ Denatured Transition of Ribonuclease A at pH* 2.8[a]

Solvent (v/v)	T_m *(°C)*
Aqueous	40.0 ± 0.5
Ethanol	
15%	38.5 ± 0.5
30%	31.5 ± 0.5
45%	20.5 ± 0.5
60%	9.0 ± 0.5
Methanol	
70%	15.0 ± 1.0

[a] Determined by absorbance change at 286 nm (*3*).

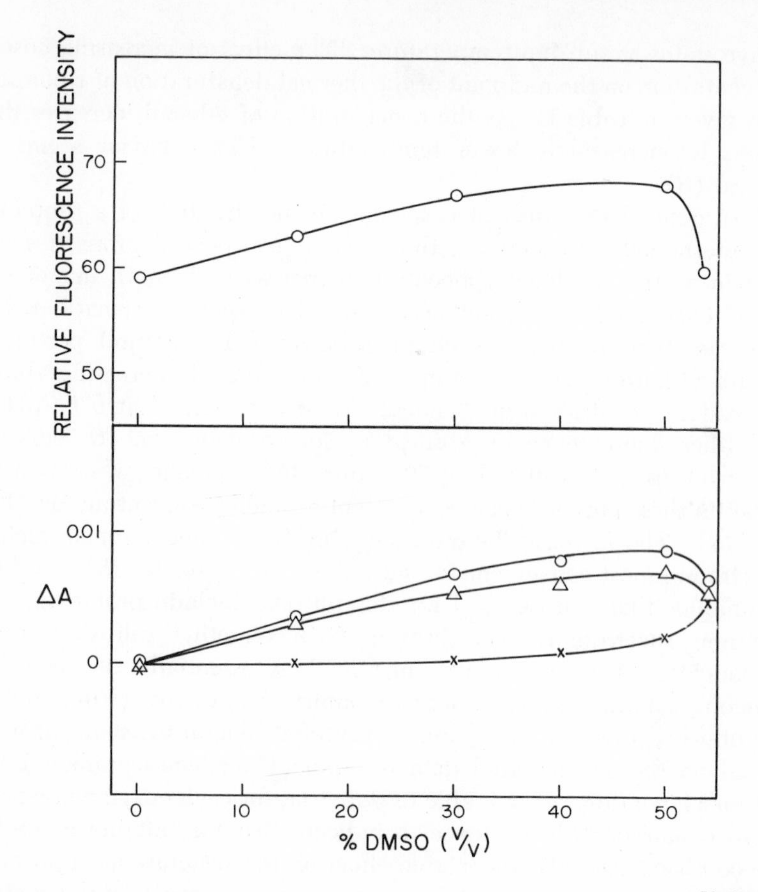

Figure 1. The effect of dimethyl sulfoxide on the structure of β-galactosidase, as monitored by: (A) the intrinsic fluorescence; (B) the intrinsic uv absorption. Conditions: 0°C, pH 7.0, excitation at 285 nm, x = 300 nm, △ = 290 nm, ○ = 280. (18)*

glucose oxidase. However, as noted below, temperature-induced structural isomerizations have been seen in a few cases. The above statements apply only to those enzyme-cryosolvent systems in which the enzyme is active. It must also be noted that for many enzymes only one of several cryosolvents examined has been found to be satisfactory. For example, many enzymes are particularly sensitive to methanol.

Catalytic. Typically the effects of the cryosolvent on the catalytic parameters may be determined by examining the effect of increasing concentration of the cosolvent on kinetic parameters, such as k_{cat}, K_m, and rate constants corresponding to more elementary steps in the overall reaction such as $k_{acylation}$, $k_{degalactosylation}$, as well as inhibition constants (e.g., K_i) and pH-rate profiles. In general, only the expected effects on

k_{cat} and pH dependence are found (*11, 16, 17, 18*). However, for K_m and K_i one frequently finds an exponential increase as the cosolvent concentration increases (*11, 17*). Although this was attributed to a combination of dielectric and competitive inhibition effects in earlier publications (*16, 17, 19, 20*), it is more likely due to a hydrophobic effect (*18*) in which hydrophobic substrates partition more favorably to the bulk cryosolvent than to the active site, compared with bulk water as the solvent.

Effects of Low Temperature on Structure

Since most proteins are in a delicately balanced equilibrium with solvent water, large concentrations of organic solvents might be expected to cause substantial perturbations. We attribute the demonstrated stability of enzymes and proteins in high concentrations of organic solvent at subzero temperatures to the following main features: the increase of dielectric constant with decreasing temperature (*21*); the increase of hydrogen-bond strength with decreasing temperature (*22*); opposing effects on hydrophobic interactions which tend to cancel as the temperature drops (*5*); and the "trapping" of the native state due to the high activation energy for denaturation, as well as the stabilizing influence of substrate or other bound ligands (*5*).

Because our current understanding of the interactions between water and proteins is relatively poor it is not possible to give a detailed, quantitative explanation of the additional effects and complications caused by an added organic cosolvent and subzero temperatures. It is possible, however, to make some qualitative estimates of the anticipated effects and to rationalize to some extent the observed stability under such conditions. Adverse effects on the enzyme structure could arise from destabilization of the native conformation or from stabilization of a non-native noncatalytically active (denatured) state. In this discussion the former is of prime interest. A more detailed discussion of the anticipated effects of low temperatures and organic cosolvents on protein structure has been given elsewhere (*5*).

We have observed three types of effects on the structures of enzymes as the temperature is lowered in cryosolvents: (1) no apparent change in the protein conformation, with the possible exception of decreased mobility of the surface side chains; (2) conformational transitions, usually marked by little effect on the catalytic properties; and (3) increased association of subunits. In most cases no detectable effects of decreasing temperature on the enzyme's structure have been detected by such procedures as monitoring the intrinsic fluorescence (*16*), or intrinsic visible absorbance in the case of flavin enzymes (Fink and

Ahmed, in preparation), or in the proton nmr spectrum (e.g., subtilisin) (Fink and Tsai, in preparation). In addition, the structure of crystalline elastase has been compared at 25°C in aqueous solution with that at −55°C in 70% methanol. Except for changes on the protein's surface, due to decreased mobility of surface side chains, the structures are identical (*13*).

When β-galactosidase is cooled in either dimethyl sulfoxide, methanol, or ethylene glycol–methanol cryosolvents, under certain conditions of enzyme concentration and ionic environment, an increase in the uv absorption of the enzyme is noted at temperatures below −30°C. The kinetics of this reaction are first-order and the rates are similar in all three cryosolvents (*18*). The catalytic activity remains unchanged during the progress of this reaction. The increase in absorbance is attributed to a temperature-induced conformational change in the β-galactosidase subunits, which does not affect the active-site or catalytic activity. A similar phenomenon has been noted also with α-chymotrypsin (Fink and Good, unpublished observations).

Because many metabolically significant enzymes are oligomeric it is of particular interest to determine the effects of cryosolvents and subzero temperatures on the state of association of multisubunit enzymes. At appropriately low temperatures (below the denaturation transition) we find no evidence of subunit dissociation in the case of β-galactosidase in aqueous dimethyl sulfoxide (*18*), glucose oxidase in aqueous methanol–ethylene glycol (Fink and Ahmed, in preparation), and liver alcohol dehydrogenase (LADH) in aqueous dimethyl sulfoxide or dimethyl formamide (Fink and Geeves, unpublished observations). However in the case of β-glucosidase (almond) the tetrameric enzyme dissociates at 25°C in 50% dimethyl sulfoxide, as determined by loss of catalytic activity, and gel-filtration chromatography (*24*). As the temperature decreases the degree of association increases: from 13% at 25°C, pH* 7.0 to 100% at −17°C (see Table III) (*24*). Confirmation that inactive monomers predominate at higher temperatures in this cryosolvent comes

Table III. The Effect of Temperature on the Dissociation of β-Glucosidase in 50% Dimethyl Sulfoxide[a]

Solvent (v/v)	*Temperature (°C)*	*% Tetramer[b]*
Aqueous	25	100
50% Dimethyl sulfoxide	25	13 ± 5
	0	76 ± 7
	−17	100

[a] pH* 7.0.
[b] Using exclusion chromatography on CPG-glycophase porous glass beads, 100 × 1 cm column (*24*).

Biochemistry

from measurements of the amount of *p*-nitrophenol liberated in the reaction with *p*-nitrophenyl-β-D-glucoside (*24, 29*), which indicate 100% active tetramer at −25°C and below, and only 30% tetramer at −5°C (*24*).

Effects of Low Temperature on the Rate of Catalysis

The effect of temperature on the catalytic reaction is most easily determined with Arrhenius plots. Such plots of the turnover rate, or of the rate of individual steps in the catalytic reaction (e.g., $k_{acylation}$, $k_{deglycosylation}$) can be very useful in comparisons between the catalytic reaction in the cryosolvent at subzero temperature and the reaction under normal conditions.

Typical effects of temperature on catalysis will be illustrated with the reaction of β-galactosidase and *o*- and *p*-nitrophenyl galactosides (*18*). The minimum reaction pathway may be represented by Scheme 1 in which ES is the noncovalent Michaelis complex, and EG represents an enzyme-galactose intermediate (*25, 26*).

Scheme 1

$$E + S \rightleftharpoons ES \xrightarrow{k_2} \underset{\underset{P_1}{+}}{EG} \xrightarrow{k_3} E + G$$

From kinetic considerations k_2 is rate-limiting for the *p*-nitrophenyl galactoside substrate, and k_3 is rate-determining for the *o*-nitro-substrate at 25°C, aqueous solution (*27*). For both substrates the Arrhenius plots at subzero temperatures were found to be linear (Figures 2A and B) indicating no change in the rate-limiting step, and no significant structural perturbation of the protein (*18*). The affinity of enzymes for their substrates would be expected to increase with decreasing temperature because the reaction in the forward direction is diffusion controlled, and hence has a low energy of activation compared to the dissociation reaction. Because the enzyme-substrate dissociation constant is a major component of the K_m term, one would expect that K_m, in many cases, would also decrease with decreasing temperature. This is indeed observed, for example, in the above-mentioned β-galactosidase case (*18*).

We have noticed for several systems examined that extrapolations of the Arrhenius plots for intermediate formation to ambient temperatures suggest that the rates of intermediate interconversion will be of the same order of magnitude at the temperatures at which the enzymes normally operate (*12*). The convergence of Arrhenius plots is shown for the papain-catalyzed hydrolysis of *N*-carbobenzoxy-L-lysine *p*-nitroanilide in Figure 3 (*28*).

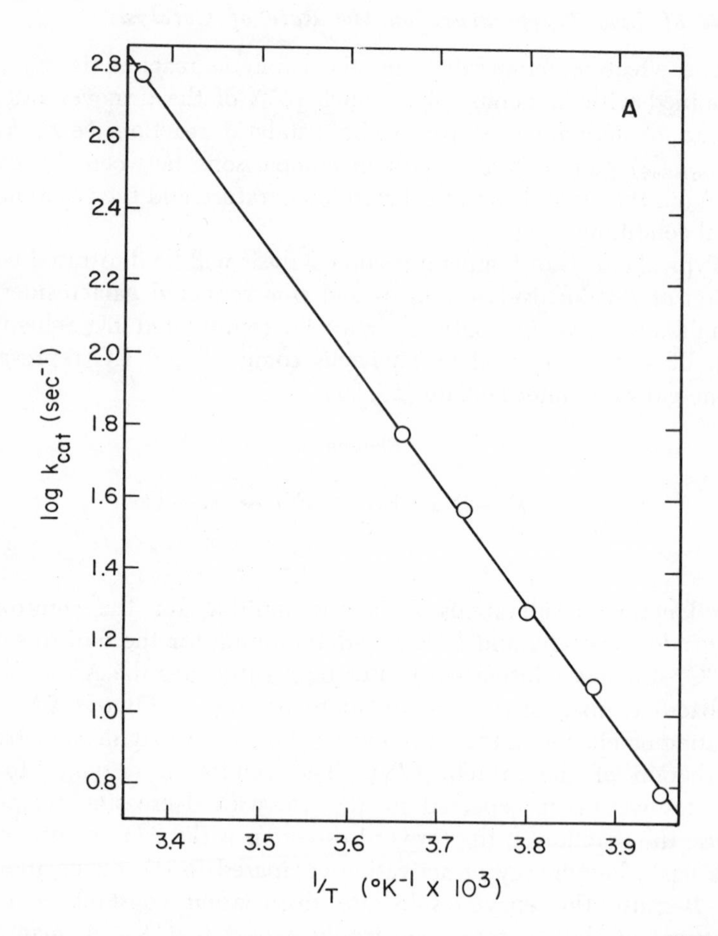

Biochemistry

Figure 2A. Arrhenius plot for the reaction of β-galactosidase with p-*nitrophenyl-β-*D*-galactoside in 60% aqueous dimethyl sulfoxide, pH* 7.0. Rate-limiting step is degalactosylation.* (18)

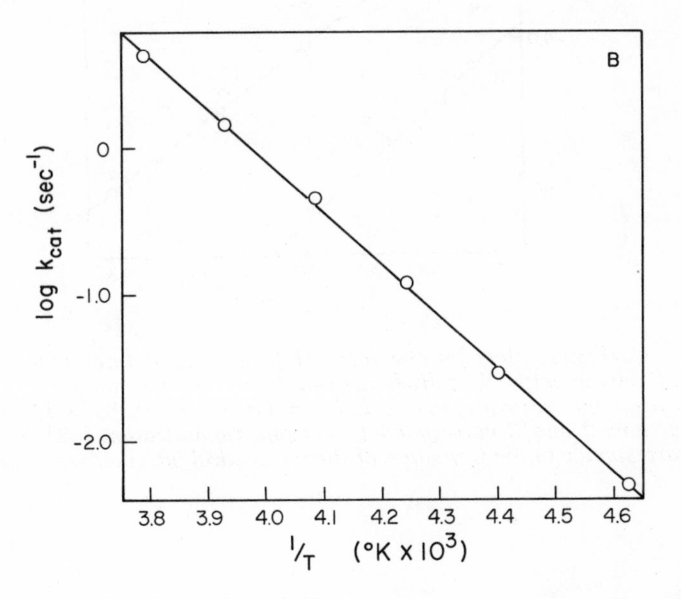

Biochemistry

Figure 2B. Arrhenius plot for the reaction of β-galactosidase with p-*nitrophenyl-β-*D-*galactoside in 60% dimethyl sulfoxide, pH* 7.0. Rate-determining step is formation of the galactose-enzyme intermediate.* (18)

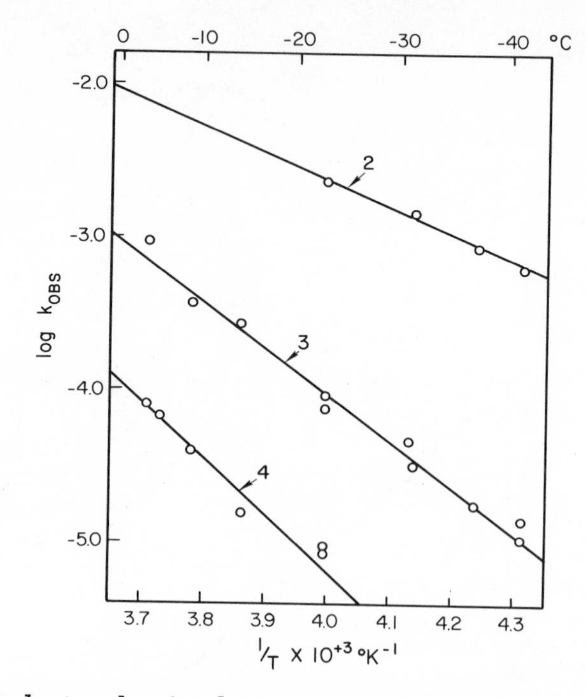

Figure 3. Arrhenius plots for the rates of formation of intermediates in the reaction of papain with N$^\alpha$*-carbobenzoxy-*L*-lysine* p*-nitroanilide. The solvent was 60% dimethyl sulfoxide, pH* 6.1,* $E_0 = 3.0 \times 10^{-6}$ M, $S_0 = 3.0 \times 10^{-5}$ M *(32). Reactions 2 and 3 correspond to enzyme isomerization (32), Reaction 4 corresponds to the formation of the tetrahedral intermediate (32).*

Table IV. Conditions Necessary

Enzyme	*Substrate*[a]	*Cryosolvent*[b]
Elastase	AcAlaProAlaPNA	70% MeOH
	CBZ-AlaPNP	70% MeOH
Trypsin	CBZ-LysPNP	65% DMSO
Papain	CBZ-LysPNA	60% DMSO
	CBZ-LysPNP	60% DMSO
Lysozyme	NAG_6	70% MeOH
β-Glucosidase	PNPGlu	50% DMSO
Subtilisin	BzPheValArgPNA	70% MeOH
Ribonuclease A	2′3′CMP	50% MeOH

[a] PNA = *p*-nitroanilide; PNP = *p*-nitrophenyl ester; NAG_6 = hexamer of *N*-acetyl-glucosamine; CBZ = N^α-carbobenzoxy-; Glu = glucoside; Bz = benzoyl; 2′3′ cyclic cytidine monophosphate.

Conditions Necessary to Stop Turnover

The temperature necessary to effectively stop the turnover reaction varies greatly depending on the particular enzyme-substrate system, the cryosolvent, the pH*, and the enzyme concentration. In some cases, where the energy of activation of the rate-determining step is large, as in the glycosidases, temperatures as high as −20°C at the pH optimum are sufficient (*24, 29, 30, 31*). By choosing a nonoptimal pH* the turnover reaction for most enzymes can be brought to a halt at −50°C or higher. Some representative data are given in Table IV. In general the often large effect of the cosolvent on K_m means that substantial additional rate reductions can be effected by working with nonsaturating substrate concentrations (*5*).

For reactions in which a chromophoric product is released part way through the catalytic reaction—for example, in protease catalysis where an acyl-enzyme intermediate is formed—it is often possible to see the release of an equivalent amount of the product formed concurrently to the formation of the enzyme–substrate intermediate. For example, as shown in Figure 4, in the reaction of papain with N^{α}-carbobenzoxy-L-lysine *p*-nitrophenyl ester in 60% dimethyl sulfoxide at pH* 6.1 (the pH optimum) a stoichiometric "burst" of *p*-nitrophenol is observed at temperatures below −40°C as the acyl-enzyme is formed, followed by no further release of *p*-nitrophenol, indicating that no turnover is occurring (*11*).

to Effectively Stop Turnover

*pH**	E_0 *(M)*	S_0 *(M)*	*Temperature (°C)*
9.2	3.8×10^{-6}	5.0×10^{-4}	−60
7.2	1.3×10^{-5}	3.2×10^{-3}	−40
7.7	2.8×10^{-5}	1.0×10^{-3}	−45
6.1	3.0×10^{-6}	3.0×10^{-5}	−15
7.0	1.4×10^{-5}	1.0×10^{-3}	−70
5.8	3.6×10^{-7}	2.0×10^{-5}	−20
7.1	4.7×10^{-5}	1.0×10^{-2}	−20
9.5	2.4×10^{-6}	6.9×10^{-5}	−21
2.1	8.0×10^{-5}	3.2×10^{-4}	−40

[b] MeOH = methanol; DMSO = dimethyl sulfoxide.

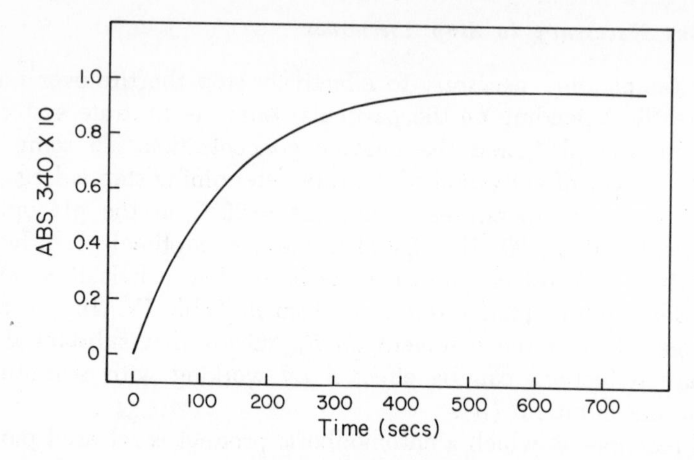

Figure 4. The acylation of papain by N^{α}*-carbobenzoxy-*L*-lysine* p*-nitrophenyl ester under nonturnover conditions in 60% dimethyl sulfoxide, −70°C, pH* 6.8.* (See *Ref.* 11)

Correspondence Between Subzero and Normal Temperature Data

In order for cryoenzymology experiments to provide mechanistically significant information, it is important that the reaction at low temperature be analogous to that under normal conditions, that is, aqueous solution and ambient temperature. One definitive way to demonstrate this is to take the kinetic data for a particular intermediate at subzero temperatures and to calculate the expected rates of formation and breakdown under normal conditions. If these rates are accessible to measurement (e.g., by stopped-flow techniques) then comparison can be made between the low-temperature and high-temperature data.

In the reaction of papain with N^{α}-carbobenzoxy-L-lysine *p*-nitroanilide, studied at subzero temperatures in 60% dimethyl sulfoxide, the slowest step preceding the overall rate-limiting step corresponds to the formation of a tetrahedral intermediate (*28*). The reaction appears as an increase in absorbance in the 360–400 nm region (Figure 5), and at high pH* (above 9) essentially all the enzyme can be trapped in the form of this intermediate. By correction for the effect of cosolvent on the rate, and extrapolation to 25°C using the Arrhenius plot, a value of 65 ± 10 s^{-1} was estimated for the observed rate of formation under a given set of enzyme and substrate conditions. The reaction should therefore be readily observable using stopped-flow spectrophotometry, as proved to be the case. The measured rate was found to be 70 ± 4 s^{-1} (*32*). In addition, at the low temperatures the concentration of the tetrahedral intermediate that could be accumulated was pH*-dependent, increasing to a maximum at high pH*. Exactly the same relationship was found in the stopped-flow studies. Similarly the rate of formation of the intermediate was pH*-dependent both at subzero and ambient temperatures (*32*).

Such correlations are not always possible because the estimated rates at 25°C may be too fast to measure with rapid-mixing techniques and because the rates of formation of more than one intermediate may be of the same order of magnitude at the higher temperature (*12*).

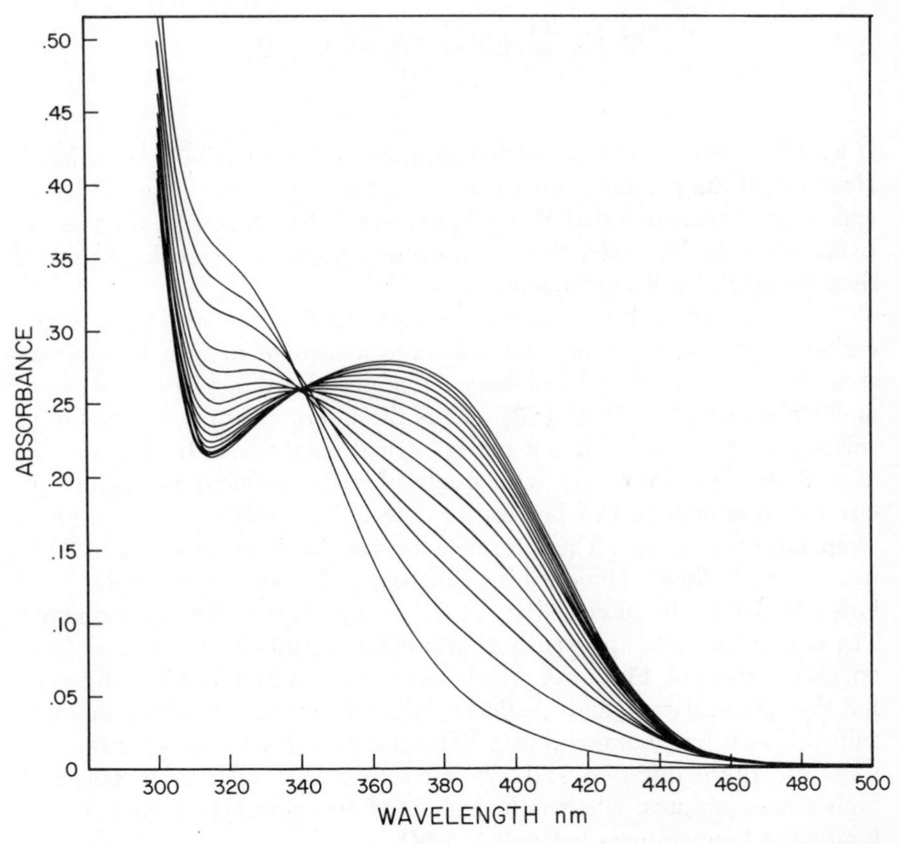

Figure 5. Formation of the tetrahedral intermediate in the reaction of papain with N^{α}*-carbobenzoxy-*L*-lysine* p*-nitroanilide in 60% dimethyl sulfoxide at −3°C, pH 9.3. The kinetics were followed by repetitive spectral scans,* λ_{max} *of the intermediate is at 361 nm. The value of* k_{obs} *is* 2.4×10^{-5} *sec*$^{-1}$ *for* $E_0 = 7.5 \times 10^{-5}$ M, $S_0 = 2.7 \times 10^{-5}$ M (*32*).

Change in Rate-Determining Step at Low Temperature

An additional potential advantage of cryoenzymology is that reactions in which intermediates cannot be detected at ambient temperature, because they breakdown more rapidly than they are formed, may undergo changes in the rate-determining step such that the intermediate breakdown becomes slower than its rate of formation at low temperatures.

The minimum reaction pathway for most hydrolytic enzymes can be represented by Scheme 2 in which ES′ is an enzyme-substrate intermediate such as an acyl-, phosphoryl-, or glycosyl-enzyme.

Scheme 2

$$\mathrm{E} + \mathrm{S} \overset{k_1}{\rightleftarrows} \mathrm{ES} \overset{k_2}{\rightarrow} \underset{\mathrm{P_1}}{\underset{+}{\mathrm{ES'}}} \overset{k_3}{\rightarrow} \mathrm{E} + \mathrm{P_2}$$

If k_2 is less than k_3 at ambient temperatures ES′ cannot be accumulated. However, if the energies of activation (E_a) of steps 2 and 3 are different and in particular such that E_a for k_3 is greater than E_a for k_2, then as the temperature is decreased there will come a point at which k_2 is greater than k_3 and ES′ will accumulate.

At least two such examples are known. In the reaction of pancreatic carboxypeptidase with an ester substrate a change in rate-determining step from formation to breakdown of a postulated acyl–enzyme intermediate occurs at −10°C (*33*). Thus the intermediate is detectable at subzero temperatures but not under normal conditions. In the reaction of β-glucosidase (almond) with *p*-nitrophenyl-β-D-glucoside a glucosyl-enzyme intermediate can be readily detected and stabilized at subzero temperatures (*24, 29*). The intermediate has also been detected at 20°C using stopped-flow techniques by Takahasi (*34*). However, Legler (*35*) failed to detect the intermediate at 37° using stopped-flow experiments. The reason for these apparently contradictory results becomes clear from an examination of Figure 6, which shows the Arrhenius plots obtained for the glucosylation and deglucosylation reactions in 50% dimethyl sulfoxide at subzero temperatures. The energies of activation for the two steps are quite different and are such that glucosylation is faster at higher temperatures, whereas hydrolysis of the glucosyl-enzyme is rate-limiting at temperatures below 0°C (*24*).

Temperature-Dependent Changes in ΔH

Although the Arrhenius plots at subzero temperatures for many of the reactions investigated have been linear (e.g., Figures 2 and 3), some cases have been observed in which changes in the enthalpy of activation

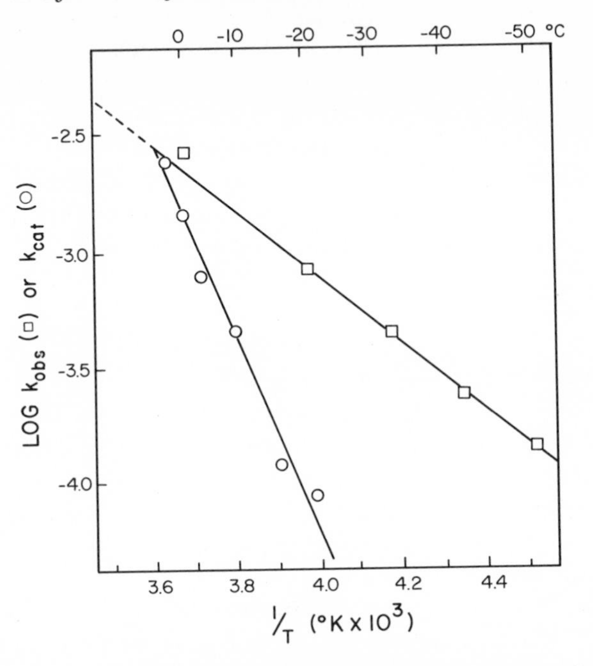

Figure 6. Change in the rate-determining step in the reaction of β-glucosidase with p-*nitrophenyl-β-*D-*glucoside in 50% dimethyl sulfoxide, pH* 7.1. The rate constant* k_{obs} *is that for formation of the enzyme-glucose intermediate;* k_{cat} *corresponds to the rate constant for deglucosylation at temperatures below 0°C.*

occur (e.g., Figure 6). Lysozyme (hen egg-white) is another such case, and represents some interesting features (*31*). Aqueous methanol cryosolvents are suitable for low temperature investigations of lysozyme catalysis (*31*). The interaction between lysozyme and the hexasaccharide of *N*-acetyl-glucosamine, or the corresponding trisaccharide inhibitor, can be followed by changes in the intrinsic fluorescence of the enzyme. Three intermediate interconversions can be detected with the substrate prior to the rate-limiting step, and two interconversions can be seen in the case of the inhibitor (*31*). Both the rate constants for each individual transformation and the dissociation constants for the overall nonturnover process have been determined as a function of temperature, both at subzero temperatures (*31*) and at higher temperatures under normal conditions (*36*). Comparison of the two sets of data indicate that a break occurs in the Arrhenius and Van't Hoff plots in the vicinity of 15–20°C. Lysozyme is known to undergo a structural rearrangement in this temperature region (*37*).

Some other systems in which nonlinear Arrhenius plots have been found at subzero temperatures have been reported by Douzou (*2*). There is no reason at present to assume that the observed changes in energy of activation at temperatures above $-100°C$ are not due to the same phenomena as cause such deviations at higher temperatures in aqueous solutions (e.g., change in rate-determining step and temperature-induced conformational change in the protein). However, at temperatures closer to absolute zero Frauenfelder's observations regarding the kinetics of ligand recombination with heme proteins suggest that adherence to the simple Arrhenius expression is no longer valid (*38*).

Conclusions

From our investigations of several different enzyme-catalyzed reactions at subzero temperatures in aqueous organic solvent systems we have been able to make a number of generalizations. The first is that, in many cases, the reaction pathway and catalytic mechanism seem to be unchanged under these conditions, compared to the reaction under normal conditions. These conclusions are based on a variety of evidence, including kinetic and thermodynamic properties, types of intermediate detectable, and lack of structural changes in the protein. Secondly, it appears that in most, if not all, cases the initial binding of substrate to the enzyme is followed by an isomerization step involving at least some of the active-site catalytic groups. Furthermore it seems, based on a limited number of systems, that the extrapolated rates for each intermediate transformation to 25–37° indicates that under normal conditions the rates of most of the intermediate interconversions are very similar. This is entirely consistent with the ideas of enzyme evolution (catalytic) as enunciated by Knowles and coworkers (*39*). Implicit in the above findings is the fact that the structure of the enzyme at subzero temperatures, in the appropriate cryosolvent, is not significantly different from that in aqueous solution and ambient temperatures.

Although the enzymes used in this article to illustrate various features of cryoenzymology have been relatively simple and well characterized, the rather general applicability of the technique to protein–ligand interactions is indicated by successful investigations of much more complicated systems. Several biochemical systems involving electron transport have been subjected to detailed examination at subzero temperatures in fluid solvents. For example, chloroplasts and some of their individual components in the photosynthetic electron-transport chain have been found by Cox (*40, 41*) to be functional, indicating that cryoenzymological techniques are applicable to membrane-bound enzymes. In fact Douzou and coworkers have carried out extensive investigations of liver microsomal, and bacterial, oxidase (hydroxylating) systems involving cytochrome P_{450} (*42–44*).

Electron transport in mitochondrial systems has been studied by Chance in some low-temperature (to 77°K) experiments involving intermediates in the oxidation of cytochrome oxidase (*45, 46*). Flash photolysis was used to dissociate the CO complex under conditions where the reassociation with oxygen is favored. Intramolecule enzyme-like reactions, which can be carried out in nonfluid (i.e. glass) solvents have also been probed successfully at very low temperatures (e.g., rhodhopsin and bacteriorhodopsin (*47*) and the reassociation of ligands to heme proteins in which Fraeunfelder and coworkers (*38, 48, 49*) have observed some very intriguing phenomena over the temperature range 20 to 300°K). One other investigation of note is that of Hastings and coworkers (*50, 51*) in which several intermediates in the reaction of bacterial luciferase have been trapped and characterized. Reviews of these investigations may be found elsewhere (*2, 8*).

One of the ultimate goals of our cryoenzymology studies is to obtain detailed, high-resolution structural information about each intermediate on the reaction pathway. The present method of choice to accomplish this goal is X-ray crystallography. Other techniques can provide specific information in the detail required for only limited parts of the structure at best. The first steps in this direction have now been accomplished. For example, the feasibility of obtaining X-ray diffraction data on trapped crystalline enzyme–substrate intermediates has been demonstrated (*13, 23*) as follows. Crystals of elastase were grown in the normal manner, transferred to 70% methanol cryosolvent, and the substrate allowed to diffuse in at −55°C until the acyl-enzyme was formed (in $\geqslant$ 80% yield). Diffraction data were then collected, leading to the eventual solution of the structure of the trapped acyl-enzyme (*13*). Based on experiments underway we expect that a set of "time-lapse" pictures of the step-wise transformation of substrate to product will become available in the near future. Clearly the potential of cryoenzymology is just beginning to become apparent.

Literature Cited

1. Freed, S. *Science* **1965**, *150*, 576.
2. Douzou, P. "Cryobiochemistry"; Academic: Paris, 1977.
3. Fink, A. L.; Gray, B. L. In "Biomolecular Structure and Function," Agris, P. F., Sykes, B., Loeppky, R., Eds.; Academic: New York, 1978, p 471.
4. Fink, A. L. *Acc. Chem. Res.* **1977**, *10*, 233.
5. Fink, A. L. *J. Theor. Biol.* **1976**, *61*, 419.
6. Fink, A. L.; Geeves, M. "Methods in Enzymol"; Academic: New York, 1979.
7. Makinent, M. W.; Fink, A. L. *Ann. Rev. Biophys. Bioeng.* **1977**, *6*, 301.
8. Douzou, P. *Adv. Enzymol.* **1977**, *45*, 157.
9. Douzou, P. *Methods Biochem. Anal.* **1974**, *22*, 401.
10. Douzou, P. *Mol. Cell Biochem.* **1973**, *1*, 15.

11. Fink, A. L.; Angelides, K. J. *Biochemistry* **1976**, *15*, 5287.
12. Fink, A. L. *Biochemistry* **1976**, *15*, 1580.
13. Alber, T.; Petsko, G. A.; Tsernoglou, D. *Nature* **1976**, *263*, 297.
14. Douzou, P.; Hui Bon Hoa, G.; Maurel, P.; Travers, F. In "Handbook of Biochemistry and Molecular Biology," Fasman, G., Ed.; Chemical Rubber Co: Cleveland, Ohio, 1976.
15. Hui Bon Hoa, G.; Douzou, P. *J. Biol. Chem.* **1973**, *248*, 4649.
16. Fink, A. L. *Biochemistry* **1973**, *12*, 1736.
17. Fink, A. L. *Biochemistry* **1974**, *13*, 277.
18. Fink, A. L.; Magnusdottir, K. *Biochemistry* **1979**, submitted.
19. Clement, G. E.; Bender, M. L. *Biochemistry* **1963**, *2*, 836.
20. Mares-Guia, M.; Figueiredo, A. F. S. *Biochemistry* **1972**, *11*, 2091.
21. Akerlöf, G. *J. Am. Chem. Soc.* **1932**, *54*, 4125.
22. Kavanau, J. L. *J. Gen. Physiol.* **1950**, *34*, 193.
23. Fink, A. L.; Ahmed, A. I. *Nature* **1976**, *263*, 294.
24. Fink, A. L.; Weber, J. *Biochemistry* **1979**, submitted.
25. Fink, A. L.; Angelides, K. J. *Biochem. Biophys. Res. Commun.* **1975**, *64*, 701.
26. Wallenfels, K.; Weil, R. *The Enzymes* **1972**, *7*, 618.
27. Sinnott, M. L.; Souchard, I. J. L. *Biochem. J.* **1973**, *133*, 89.
28. Angelides, K. J.; Fink, A. L. *Biochemistry* **1978**, submitted.
29. Fink, A. L.; Good, N. *Biochem. Biophys. Res. Commun.* **1974**, *58*, 126.
30. Hui Bon Hoa, G.; Douzou, P.; Petsko, G. A. *J. Mol. Biol.* **1975**, *96*, 367.
31. Fink, A. L.; Homer, R.; Weber, J. *Biochemistry* **1979**, submitted.
32. Angelides, K. J.; Fink, A. L. *Biochemistry* **1979**, in press.
33. Makinen, M. W.; Yamamura, G.; Kaiser, E. T. *Proc. Nat. Acad. Sci. U.S.* **1976**, *73*, 3882.
34. Takahashi, K. *J. Sci. Hiroshima Univ. Ser. A. Phys. Chem.* **1975**, *39*, 237.
35. Legler, G. *Acta Microbiol. Acad. Sci. Hung.* **1975**, *22*, 403.
36. Banerjee, S. K.; Holler, E.; Hess, G. P.; Rupley, J. A. *J. Biol. Chem.* **1975**, *250*, 4355.
37. Saint-Blanchard, J.; Clochard, A.; Cozzone, P.; Berthou, J.; Jollès, P. *Biochim. Biophys. Acta* **1977**, *491*, 354.
38. Alberding, N.; Austin, R. H.; Chan, S. S.; Eisenstein, L.; Frauenfelder, H.; Gunsalus, I. C.; Nordlund, T. M. *J. Chem. Phys.* **1976**, *65*, 5631.
39. Albery, W. J.; Knowles, J. R. *Biochemistry* **1976**, *15*, 5631.
40. Cox, R. *Eur. J. Biochem.* **1975**, *55*, 625.
41. Cox, R. *Biochim. Biophys. Acta* **1975**, *387*, 588.
42. Debey, P.; Hui Bon Hoa, G.; Douzou, P. *FEBS Lett.* **1973**, *70*, 2633.
43. Debey, P.; Balny, C.; Douzou, P. *Proc. Nat. Acad. Sci. U.S.* **1973**, *70*, 2633.
44. Debey, P.; Balny, C.; Douzou, P. *FEBS Lett.* **1973**, *35*, 86.
45. Chance, B.; Graham, N.; Legallais, V. *Anal. Biochem.* **1975**, *67*, 552.
46. Chance, B.; Saronio, C.; Leigh, J. S. *J. Biol. Chem.* **1975**, *250*, 9226.
47. Hess, B.; Oesterhelt, D. In "Dynamics of Energy-Translucing Membranes", Ernster, Estabrook and Slater, Eds. Elsevier: Amsterdam, 1974; p 257.
48. Austin, R. H.; Beeson, K. W.; Eisenstein, L.; Frauenfelder, H.; Gunsalus, I. C. *Biochemistry* **1975**, *14*, 5355.
49. Austin, R. H.; Alberding, N.; Beeson, K.; Chan, S.; Eisenstein, L.; Frauenfelder, H.; Gunsalus, I. C.; Nordlund, T. *Croat. Chem. Acta* **1977**, *49*, 287.
50. Hastings, J. W.; Balny, C. *J. Biol. Chem.* **1975**, *250*, 7288.
51. Balny, C.; Hastings, J. W. *Biochemistry* **1975**, *14*, 4719.

RECEIVED June 16, 1978.

4

Parameters of Freezing Damage to Enzymes

WILLIAM N. FISHBEIN and JOHN W. WINKERT

Biochemistry Division, Armed Forces Institute of Pathology,
Washington, DC 20306

An extensive evaluation of the several parameters involved in freezing damage to catalase has revealed a number of analogies to cellular systems, although the operative mechanisms differ. In all media, damage increased at slower warming rates and on progressive dilution of the enzyme. In phosphate buffer, it increased with faster freezing rates and with lower nucleation temperatures. In phosphate-buffered saline it also increased at very slow freezing rates, thus generating an optimum recovery at cooling rates of 1–20°/min. Here damage was progressive at −40° and was aggravated by low doses of common cryoprotectants, although not by polymers or oligosaccharides. Moreover, the sugars blocked the damaging effects of the cryoprotectants which in turn blocked the protective effects of the polymers, suggesting a hierarchy of biologic interactions by these agents. Several of these features have also been demonstrated with another enzyme, adenylate deaminase. Most of the findings can be explained by pH changes during freeze-thaw and by nonequilibrium phase transitions.

Most enzymes can be stored with impunity in an ordinary freezer with little regard for cooling and warming rates, seeding, storage temperature, and the other parameters known to be of importance in the storage of cells and tissues. This relative simplicity of storage in the usual case has resulted in a paucity of detailed studies of freezing damage to enzymes. We have undertaken such a study with the enzyme catalase from the perspective of the cryobiologist. By careful consideration of all of the factors which have been found to be important in complicated cellular systems, we are seeking what behavior may be observed in the simplest possible biologic system, an isolated macromolecule. The generality of the observations to follow is uncertain, since only catalase has been studied in detail; however, we will mention other studies as available, at appropriate points.

The materials, methodology, and instrumentation used in this work have been detailed elsewhere (*1, 2*) and will be repeated here only when essential to the understanding of specific experiments. The catalase, from Sigma Chemical Co., showed a single band on acrylamide gel electrophoresis with an estimated molecular weight of 250,000 (indicating persistence of tetrameric structure) at the lowest concentration tested, 1 μg/mL. All cooling and warming rates up to 30°/min. were linearly controlled with the apparatus previously described (*1*); higher rates are average values for the range specified.

Studies in Neutral Potassium Phosphate

We began this study using catalase diluted in 10m*M* neutral potassium phosphate buffer. Equilibrium phase diagrams for this buffer show a eutectic at −17° and no more than 0.5 units pH change throughout the freezing range (*3*). This allows us to presume that any damage occurring from freeze-thaw is not due to alterations in pH. Using a standard freeze–thaw procedure, we first evaluated the sensitivity of the enzyme to damage at different concentrations as shown in Figure 1.

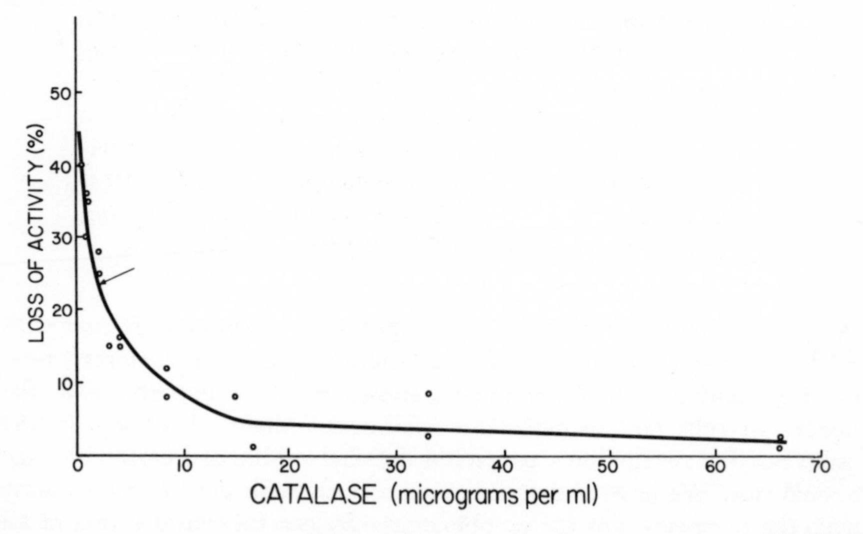

Figure 1. Effect of enzyme concentration on freeze–thaw inactivation of catalase. Catalase was frozen unseeded at 15°C/min to −78° and warmed at 10°C/min to +5.0°C (1).

Sensitivity to damage markedly increased with catalase dilution as has been observed for a number of other enzymes (*4–11*). Although not shown in Figure 1, we have tested concentrations up to 1 mg/mL without observing any return of freeze-thaw damage; so sensitivity appears to increase monotonically with enzyme dilution. We selected a concentration of 1.7 μg/mL, as shown by the arrow, for further studies, since this level displayed an appreciable and reproducible activity loss and was convenient for assay. To insure reproducibility of the system the following additional factors were then evaluated, and found to have no influence on the freezing damage: increase in buffer concentration to 100m*M*, freezing nadir of $-30°$ to $-196°$, storage at $-30°$ or below for 10 minutes to 24 hours, thawing time at $+5°$ of 1/5 to 10 minutes, and post-thaw standing at 0° of 10 to 120 minutes.

The next question was the nature of the inactivation. A number of enzymes have been shown to be susceptible to cold inactivation (*12–18*), or what the cryobiologist might term thermal shock (*19, 20*), resulting from a fall in temperature unaccompanied by any phase changes. This phenomenon is restricted primarily to large oligomers, and results from dissociation, or rarely, from conformational unfolding (*17*). Thermodynamic reversibility has been demonstrated in the case of two self-associating proteins (non-enzymatic), tobacco mosaic virus coat protein (*21*), and calf-brain tubulin (*22, 23*). Detailed physicochemical analyses have been presented for these entropic reactions. Of course the same phenomenon might also occur on freezing of other macromolecules, consequent to a still greater fall in temperature plus the attendant increase in solute levels; analysis, however, would be much more difficult in the presence of phase changes.

We interpret the following experiments to indicate that catalase inactivation is irreversible, associated with denaturation, and is true freezing damage not cold inactivation. First, our attempts to produce damage by dropping or cycling the temperature without freezing have been unsuccessful (some of these will be mentioned below). Second, post-thaw activity remained unchanged after 24 hours standing at 0°. Third, acrylamide gel electrophoresis of the enzyme taken through five freeze–thaw cycles to produce 85% damage, showed a polydisperse smear replacing the normal band of molecular weight 250,000. No subunit bands were present, although dissociation and reaggregation might have generated the polydispersity.

The next factor investigated was the warming rate, with and without prior seeding by air injection of microsamples of frozen buffer at $-1°$ to $-2°$. For completeness, Figure 2 shows the results in samples containing added KCl and added NaCl as well as in phosphate buffer alone. The patterns are essentially the same, although there is greater damage in

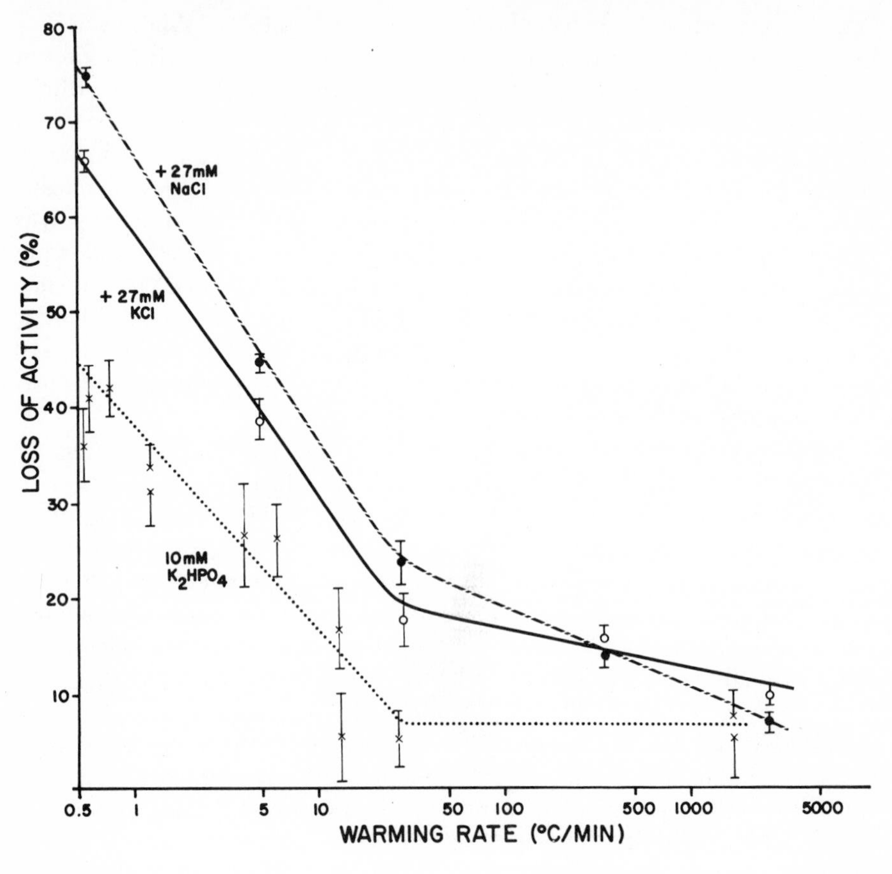

Figure 2. Warming-rate dependence of catalase inactivation in various solutions after seeding and quenching in liquid nitrogen. Mean and SE are shown for 4–6 samples in each case. Rates were controlled from −20, −30, and −50°C or lower for solutions containing phosphate only, KCl, and NaCl, respectively (20).

the presence of added salt. The pattern was also the same whether or not the samples had been seeded. Damage increased progressively as the warming rate was decreased below 20°/min; and throughout our studies this has remained the most important extrinsic factor in freeze–thaw damage to catalase.

Since warming rates in excess of 20°/min prevented enzyme damage, we utilized this feature to evaluate the temperature range over which damage occurred. In Figure 3A samples were quickly frozen at −78° then transferred to a −20° alcohol bath which was then warmed slowly to various subzero temperatures, after which the tubes were quickly immersed and swirled in a 37° water bath. The resulting warming rate is about 50°/min. Damage was negligible when slow-warming was terminated at −10° and progressed to 60% when it was continued to +2°. In Figure 3B the approach was reversed with slow warming initiated at some subzero temperature and continued until thawing was complete. This experiment is less exact because the transition from rapid to slow warming occurs gradually in the damage zone, but it does show that 60% damage was the maximum obtainable whether slow warming was begun (nominally) at −5°, −10°, or −20°. We can, therefore, conclude that the damage zone for catalase in this medium extends no lower than −7° or −8°.

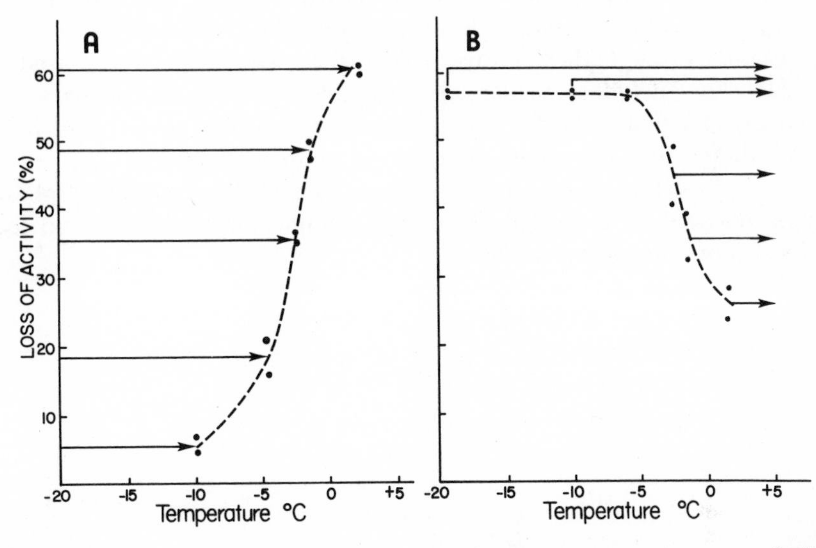

Figure 3. Temperature zone of freeze–thaw damage to catalase. Half-milliliter aliquots of the enzyme (1.7 μg/ml) were frozen at 900°C/min without seeding and held at −78°C.

In both graphs, the arrows show the temperature range over which slow warming (0.6°C/min) was carried out, and the dashed lines show the inactivation signature. (A) The tubes were transferred to a −20°C bath (warming rate ≃ 50°C/min) and after equilibration were warmed at 0.6°C/min to various temperatures, after which pairs of tubes were rapidly immersed and swirled in a 30°C water bath (warming rate 30–50°C/min). Negligible inactivation occurred when slow warming was interrupted at −10°C. Interruption at higher temperatures resulted in progressive inactivation to a maximum of 60% when slow warming was continued to +2°C. (B) Pairs of tubes were transferred to a bath at various temperatures between −20 and +2°C and subsequently warmed to +5°C at 0.6°C/min. Average warming rate on transfer was ≃ 50°C/min but this dropped sharply as the endpoint was approached, so that rates less than 5°C/min were present for several degrees below the indicated transfer temperature. The maximum inactivation obtained was 60%, whether slow warming was initiated at +20, −10, or −5°C. Initiation at higher temperatures resulted in progressively less damage (1).

The effect of cooling rate was next investigated using a warming rate of 0.6°/min. In the absence of seeding, a constant level of damage was obtained at all cooling rates tested. We noted, however, that supercooling increased progressively at lower cooling rates with spontaneous nucleation occurring at progressively lower temperatures. When this factor was eliminated by seeding all solutions at − 2°, a clear cut dependence of damage on the cooling rate was observed as shown in Figure 4. Damage increased with cooling rate up to 5°/min after which it plateaued. This indicated that the rate of freezing, i.e., solidification, was a significant factor in damage but was obscured in the absence of seeding by the progressive supercooling and lower nucleation temperatures occurring at low cooling rates.

This was verified by the experiment shown in Figure 5 where seeding was carried out at progressively lower temperatures. Although all samples were carried through the identical freeze–thaw procedure, those seeded at − 11° showed twice the damage sustained by those seeded at − 1°. Moreover, less than 8% damage appeared in samples seeded and held at − 1° then slowly rewarmed, and also in samples supercooled at − 11° and slowly rewarmed without freezing. This indicates that a phase change followed by essentially complete solidification is involved in the production of damage to catalase.

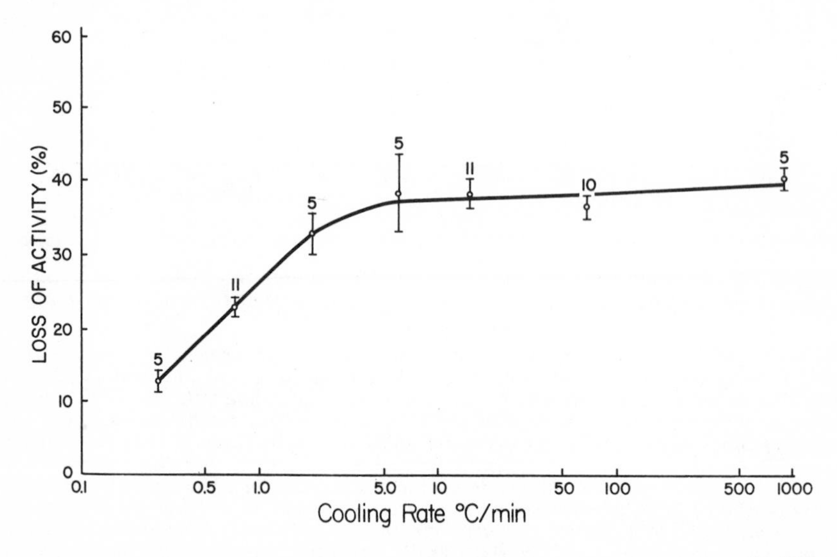

Cryobiology

Figure 4. Cooling rate dependence of freeze–thaw inactivation of seeded catalase solutions. After seeding at −2°C, samples were cooled at the rate noted to −25°C or below, held at −78°C, then warmed at 0.6°C/min from −25 to +5°C (1).

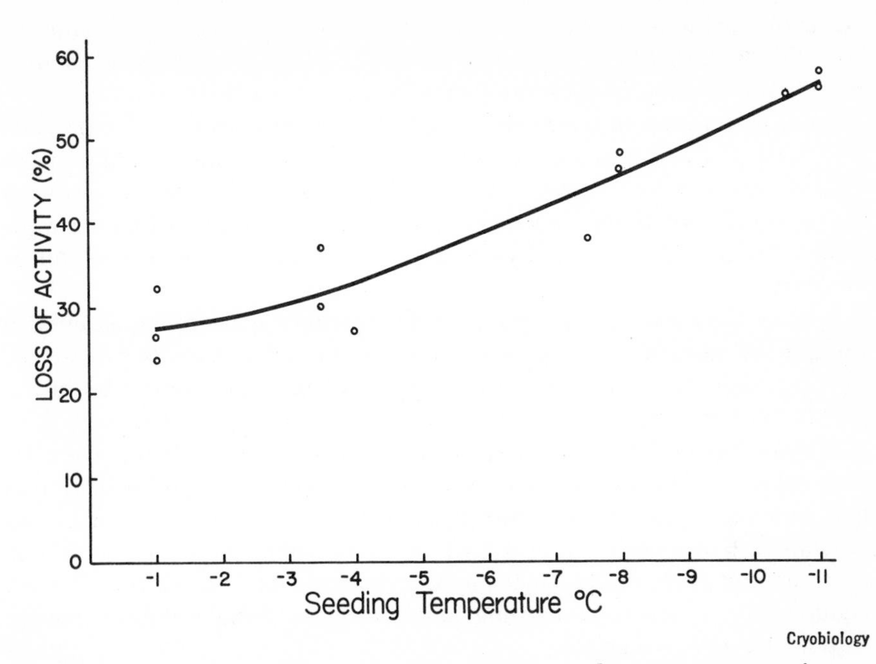

Figure 5. Effect of seeding temperature on freeze–thaw inactivation of catalase. All samples were cooled at 0.8°C/min to the seeding temperature noted, and thence to −20°C, followed by warming at 0.7°C/min to +4°C. The solid line is least-squares best-fitting parabola (1).

The possibility of thermal shock attending the increasing concentration of phosphate on freezing was investigated by incubating catalase in a saturated solution of neutral potassium phosphate at room temperature (2.2*M*) and also at 2° where a considerable amount of phosphate precipitated. Table I shows that some inactivation did occur in stationary samples but did not differ significantly at 2° and 23°. In stirred samples, however, damage was severe and significantly worse at 23°, where no precipitate was present and therefore, enzyme adsorption and pH changes could not have contributed to activity loss. Since inactivation does not

Table I. Percent Recovery of Activity[a]

	2°	*23°*
Stationary	74.4 ± 5.5	81.7 ± 4.2
Magnetic stirring	34.1 ± 1.1	18.5 ± 1.0

[a] Activity recovered from solutions of 1.7 μg catalase/mL 2.2*M* potassium phosphate, pH 7.0 after 2.5 hrs incubation at two temperatures, with or without gentle magnetic stirring, using a large external water bath to maintain constant temperatures. Mean ± SE are shown for 3 tubes in each experiment. The solution was clear at 23°, but a heavy precipitate was present at 2°. Supernatant aliquots were taken directly from the incubating tubes for assay at 23° in all cases.

occur on stirring solutions of catalase in dilute potassium phosphate, we conclude first, that high phosphate levels can produce significant inactivation of the enzyme in solution, probably due to the high ionic strength; second, that a drop in temperature makes no contribution to this process; and third, that shearing or interfacial forces between the protein and the solution make a large contribution to this process. Stirring is not used during our freeze–thaw procedures but these results suggest that attempts to increase the thawing rate by stirring may introduce another component of damage.

Since the sensitivity of catalase to freeze–thaw damage increases with dilution, it seemed reasonable to presume that the damaging effect of rapid solidification was due to the trapping of the macromolecules in the matrix at low concentrations, whereas slow cooling would permit their concentration and thus render them resistant to damage. If this were so, then sufficiently rapid cooling rates might be expected to produce equivalent damage at higher concentrations of catalase. We were unable to demonstrate this, however, over an 8–fold range of catalase concentrations with cooling rates up to 1000°/min. Although damage increased with cooling rate, it remained a strong function of the initial catalase concentration at every cooling rate.

We must emphasize at this point, the dominant effect of warming rate as compared to cooling rate in the production of damage to catalase. Although the cooling rate pattern shown is clear cut, it was eliminated if the warming rate was rapid, whereupon no damage is evident. In contrast, a slow warming rate elicited damage regardless of the prior cooling rate.

Although extensive studies in sodium phosphate were not undertaken, we did try, without success, to duplicate Shikama's studies on catalase in 10m*M* neutral sodium phosphate (*6*). He found a constant loss of 20% activity on quenching samples for 10 min or longer at any temperature from $-10°$ to $-80°$, even if this followed prefreezing at lower temperatures which did not cause damage. We are unable to account satisfactorily for this discrepancy.

Studies in Media Containing Buffer and Salts

The rate pattern of the damage, namely, increasing with cooling rate and decreasing with warming rate was unexpected and would be considered by most cryobiologists to indicate intracellular freezing. Now that interpretation is not so peculiar as it sounds because the enzyme is indeed exposed to any ice that is formed, just as though both were present within a cell. However, it is also obvious that none of the usual compartmental explanations for this pattern in cells, such as osmotic or membrane damage, can be invoked in this system. The cryobiologist would argue that damage due to concentrating solutes during freezing should instead produce damage that decreases with increasing cooling rate. Accepting that argument, we would say that the damage to catalase must not be due to concentrating solute, and further, upon adding this factor (by adding a neutral salt to the system) we ought to be able to induce an optimum cooling rate curve similar to those considered typical of cellular systems (*24*). We tried additions of 27m*M* KCl and NaCl to our standard potassium phosphate buffer. The equilibrium phase diagrams for these systems have also been described and indicate, first, that they should both be fully solidified by −23°; and second, that the NaCl system would become acid during freezing whereas the KCl system would not (*25*).

We decided to independently test the nadir of the damage zone for catalase in each system by storage at various subzero temperatures for periods of two to four days followed by rapid warming. In the potassium chloride–phosphate system the nadir for incremental damage was about −20°, in agreement with the phase diagrams. In the sodium chloride–phosphate system, however, the nadir was about −45°, much lower than expected. These nadirs defined the temperature range over which we would have to control the cooling and warming rates to investigate each medium.

The potassium chloride–phosphate medium was explored first, since it involved no pH change that might contribute to damage. Using a fixed warming rate (0.5°/min), damage was higher at all cooling rates than in phosphate alone but the increase was greater at the lower cooling rates to yield a flat response. Thus damage appeared to be independent of the cooling rate despite seeding.

We proceeded next to the sodium chloride–phosphate medium, the results of which are shown in Figure 6. These results provide an unequivocal demonstration of an optimum recovery cooling rate curve for a simple soluble protein. In the absence of salt, damage increased in this experiment as the cooling rate was increased from 1° to 25°/min and was relatively constant outside this range. On addition of 3–9m*M* NaCl, damage increased at all rates but especially so at very low rates to yield an optimum recovery or minimum activity loss at 0.5°/min. With 27m*M* NaCl, damage increased further but most strikingly at low cooling rates and the optimum became more pronounced and shifted to 5°/min. With 81m*M* NaCl, added damage became quite severe but the optimum remained distinct and shifted further to 20°/min.

As with cellular systems, we would argue that the presence of an optimum indicates two factors involved in freezing damage, one operating at high cooling rates and the other at low cooling rates. The nature of

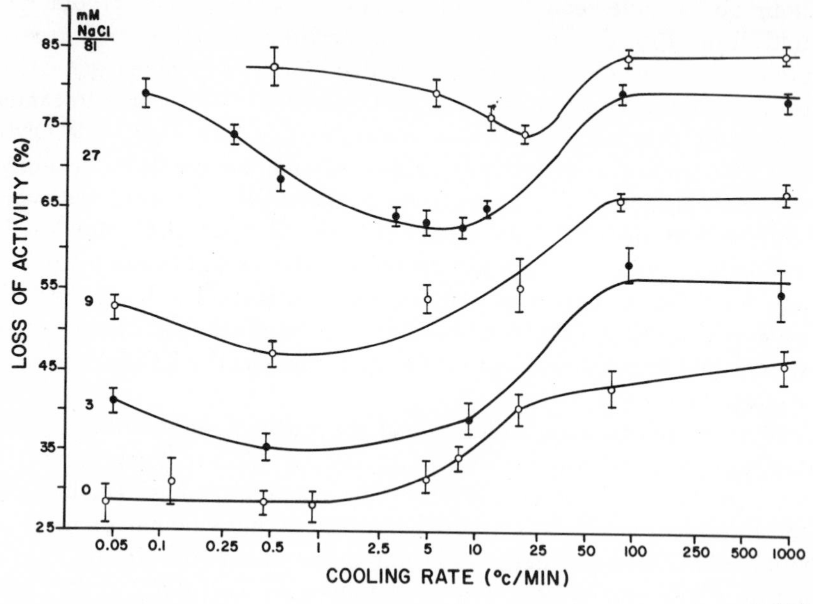

Figure 6. Cooling-rate dependence of catalase inactivation in 10mM neutral $KHPO_4$ solutions containing various concentrations of NaCl. All solutions were seeded, cooled at stated rates to −50°C or lower, and then warmed at 0.5°C/min from −50°C. Mean and SE are shown for four to six samples in each case (20).

the high cooling rate factor is still obscure but the low cooling rate factor must be salt or some consequence of salt, since we can add or remove it at will. As we freeze with higher starting levels of salt, the intermediate liquidus concentrations must increase, producing more damage at low-cooling rates and shifting the optimum or minimum to the right.

Note, however, that damage at high cooling rates is also increased in the presence of salt so that if a separate factor is involved, it is accentuated nevertheless by the presence of salt. At low cooling rates damage to catalase was much greater in NaCl than in KCl, although the eutectic compositions of the two salts differ little. The damage therefore may be due not solely to concentrating salt but to either of the additional factors operating only in NaCl solutions: acidification during freezing and the much broader temperature zone of damage.

Cryoprotectant Effects

We will first explore in more detail the unexpectedly broad damage zone, again making use of the fact that rapid warming prevents damage. A linear rate-controlled alcohol bath is warmed at 0.5°/min starting at

−70°. Seeded samples quenched in liquid nitrogen and stored at −80° are transferred to the bath at 10° intervals, warmed slowly for 10°, and then transferred to a 37° water bath for rapid thawing. Thus we can evaluate the relative effect of an added cryoprotectant in various subzero temperature zones. This consideration had become important since in preliminary experiments we had observed a net increase in damage at low, but not at higher, levels of cryoprotectants. A similar observation had been made in 1973 by Whittam and Rosano during freezing studies of α-amylase, also in phosphate-buffered saline (*26*) but their study did not address the zonal location of the damage.

Figure 7 shows the effect of adding diglyme (diethylene glycol dimethylether) to the 10m*M* phosphate–27m*M* NaCl medium. Ten m*M* diglyme was an effective protectant of catalase in all zones except that near −40°, where it produced marked enzyme damage. The weight ratio of cryoprotectant to salt is less than 1, so it should not significantly affect the eutectic temperature according to ternary phase diagram studies. Moreover the damage must not be a direct toxic effect of diglyme because a 5-fold higher dose eliminates the damage leaving instead a zone of weak protection. The same findings were obtained with the same concentration of dimethyl sulfoxide (DMSO) and of glycerol, the two most widely used cryoprotectants. Only occasionally was the damage near −40° severe enough to yield net damage in a full-range freeze-thaw experiment; in the usual case the net effect was slight protection. We emphasize, therefore, that ineffective cryoprotection may result from the interplay of two opposing effects produced by the same agent, at least in phosphate-buffered saline: protection in one temperature zone and damage in another temperature zone.

Comparison with a macromolecular cryoprotectant is shown in Figure 8. PVP (polyvinylpyrrolidone) is markedly protective in all temperature zones whereas diglyme produces marked damage in the zone near −40°. If the two agents act via the same mechanism, they should be functionally additive and a mixture should show no zone of excess damage whereas in fact diglyme produced almost as much damage in the presence of PVP as in its absence. Polyethylene glycol at molecular weights 20,000 and 4 million behaved quite analogously to PVP, while DMSO and glycerol behaved like diglyme. Thus macroprotectants were rendered inoperative in the presence of low-molecular-weight agents.

Now if the damage zone involves a persistent liquidus, as seems likely, then we should be able to simplify future experiments by quenching samples and storing them at −35° to −40° for progressive intervals before fast warming, thus yielding greater damage with less effort. We also will scale up our concentrations to maintain the same ratios while providing the 1% saline level that is used in most phase diagram studies.

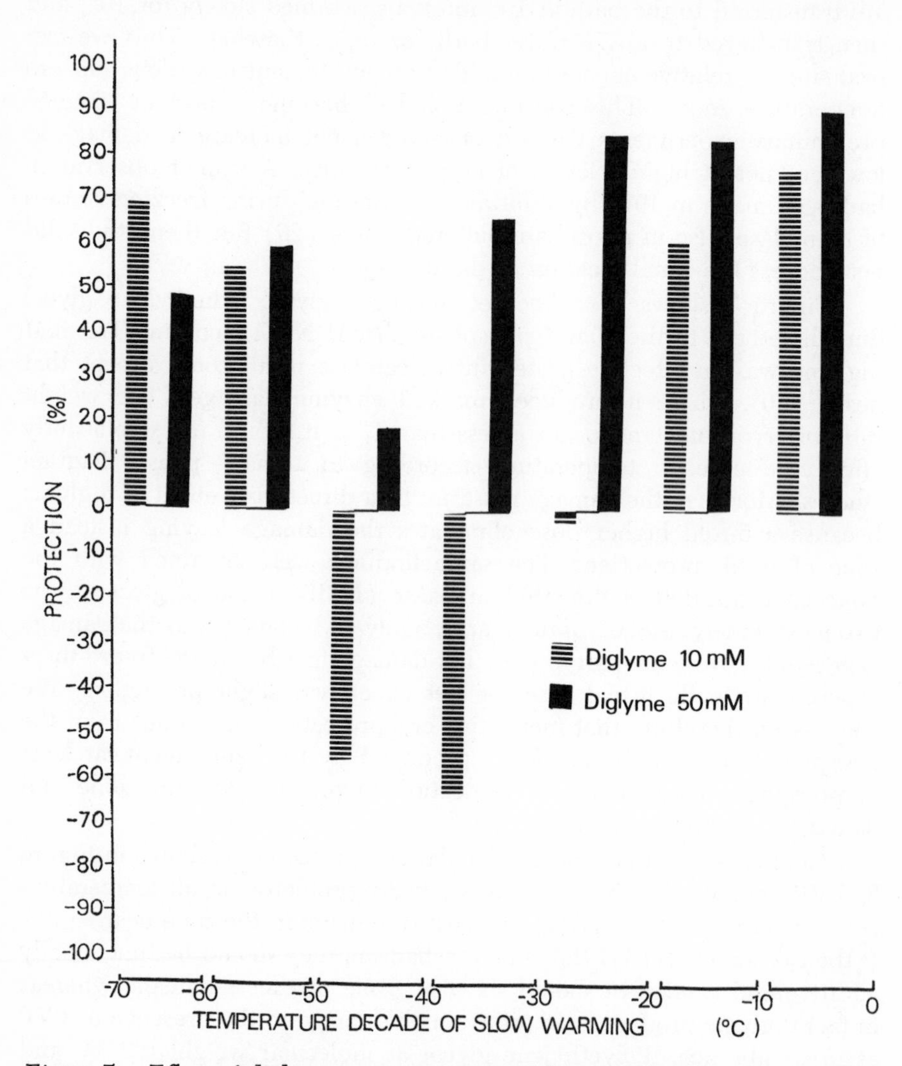

Figure 7. Effect of diglyme on recovery of catalase activity after slow warming through discrete subzero temperature zones.

Tubes contained 1.7 μg catalase/ml 10mM KPO_4 (pH 7.0) + 27mM NaCl ± 10mM or 50mM diglyme, and were seeded at −1.5° and quenched in alcohol at −80° before warming. Pairs of tubes with and without diglyme were transferred to a linear-rate warming bath (0.5°/min) at the initial temperature of each decade and after the 10° interval was traversed, they were removed and thawed rapidly in 37° water bath. The % effect was calculated as 100 (A − B)/A, where A = activity lost with diglyme absent, and B = activity lost with diglyme present.

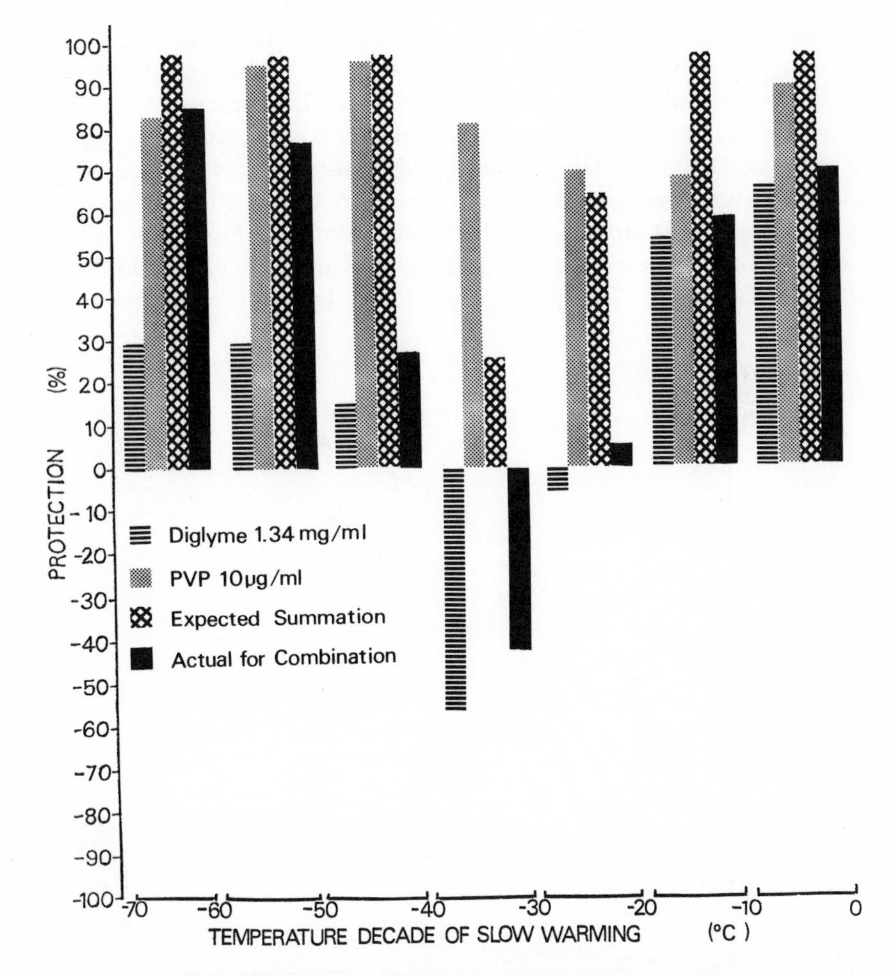

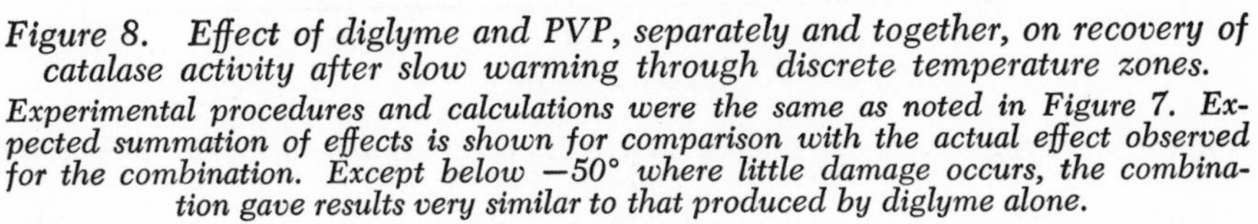

Figure 8. Effect of diglyme and PVP, separately and together, on recovery of catalase activity after slow warming through discrete temperature zones.

Experimental procedures and calculations were the same as noted in Figure 7. Expected summation of effects is shown for comparison with the actual effect observed for the combination. Except below −50° where little damage occurs, the combination gave results very similar to that produced by diglyme alone.

Figure 9 shows that on storage at −40°, catalase was not appreciably damaged in 1% saline but did show progressive damage when phosphate buffer was added and a striking further progression of damage when dimethylsulfoxide was also added. Overnight storage in this temperature range therefore could be used to evaluate a variety of other agents.

Sucrose was the first low-molecular-weight compound encountered that was protective in this zone at low concentrations and it led us to survey a variety of sugars. As shown in Table II, monosaccharides were not protective, although they produced much less damage than did glycerol (or DMSO). All oligosaccharides were protective, and no evident distinction was apparent with regard to composition or glycoside configuration. Most striking was the observation that low concentrations of sucrose could reverse the damaging effects of glycerol and DMSO as shown in Table III. The damage produced by glycerol or by DMSO

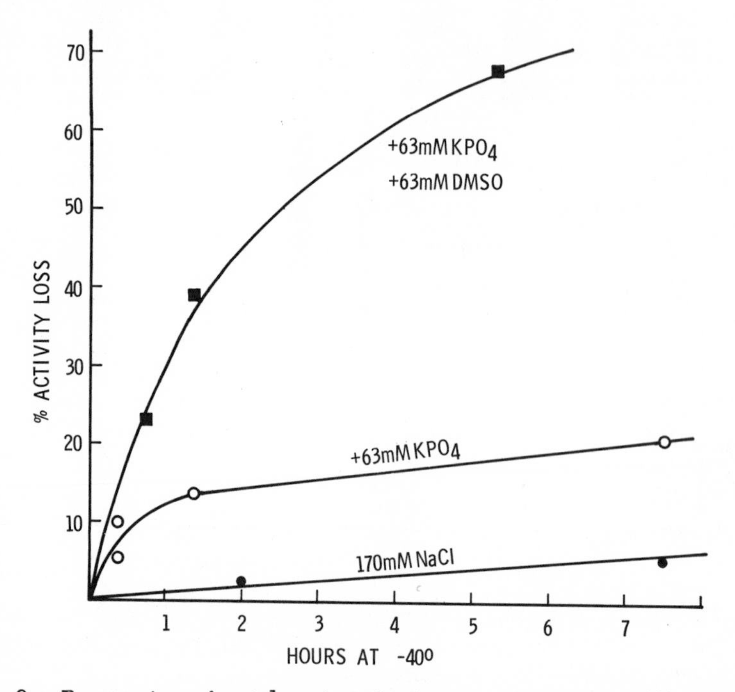

Figure 9. Progression of catalase inactivation on storage at −40° in various media.

All tubes contained 1.7 μg catalase/mL 170mM NaCl ± other solutes as noted. They were seeded and quenched at −80° before transfer to the −40° alcohol bath and thawed rapidly in a 37° water bath. Although not shown in this figure, damage in 170mM NaCl + 63mM DMSO matched that of saline alone. Freezing damage (about 20%), measured after rapid thawing without transfer to the −40° bath, was subtracted from total damage at each interval to give the storage damage shown in the figure.

Table II. Effect of Various Polyols Added to PBS[a]

Compound	Composition	Damage (−) or Protection (+) Relative to PBS Alone (%) Mean ± SE	(N)
Glycerol		−37 ± 6	(3)
Glucose	α-glu	−7 ± 3	(7)
Fructose	β-fru	−2 ± 3	(3)
Galactose	β-gal	−2 ± 1	(3)
Lactose	glu-β-gal	+ 22 ± 1	(3)
Sucrose	glu-β-fru	+49 ± 3	(15)
Cellobiose	glu-β-glu	+60 ± 3	(5)
Maltose	glu-α-glu	+63 ± 6	(10)
Melibiose	glu-α-gal	+50 ± 7	(3)
Raffinose	fru-α-glu-α-gal	+34 ± 1	(3)
Stachyose	gal-α-gal-α-glu-β-fru	+36 ± 3	(4)

[a] Added at 25m*M* to PBS (10m*M* KPO_4 + 170m*M* NaCl, pH 7) on damage to catalase (1.7 μg/mL) produced by one day storage at −36°. The % effect was calculated as 100(loss in PBS–loss with polyol)/loss in PBS. Activity loss in PBS averaged 48 ± 1% for 26 experiments. All samples were seeded and quenched to −80° before storing at −36°.

was not noticeably decreased on doubling the level from 12.5m*M* to 25m*M*. Sucrose was mildly protective at the lower level and strongly protective at the higher level, but most surprisingly it almost completely reversed the damage produced by a standard cryoprotectant when they were present together. Again this suggests that the agents act by separate mechanisms with the action of sucrose preempting that of DMSO and

Table III. Effect of Added Agents on Damage to Catalase[a, b]

	Damage (−) or Protection (+) Relative to PBS Alone (%)			
m*M Agent in PBS*	*Glycerol*	*Sucrose*	*Both*	*Expected Sum*
12.5 each	−27.4	+0.6	+2.6	−26.8
25.0 each	−30.3	+20.8	+4.8	−9.5
	DMSO	*Sucrose*	*Both*	*Expected Sum*
12.5 each	−51.2	+7.9	−7.2	−43.3
25.0 each	−48.8	+14.5	−4.8	−34.3

[a] 1.7 μg/mL catalase after one day storage at −36°.

[b] Procedure and medium are as described in Table II. However, the % effect in this table was calculated as % loss in PBS–% loss with agent(s), so as to permit determination of expected sums by simple addition. Mean values are shown for 6 tubes/group; CV was 10–15% for all groups.

glycerol. Potent cryoprotection of enzymatic function at such low concentrations of sucrose (and other oligosaccharides) has also been demonstrated in more complicated systems, such as chloroplast thylakoids (*27*) and liver mitochondria (*28*).

Freezing Damage to Adenylate Deaminase

The damage zone near −40° presents another curious parallel for cryobiologists. Biologic systems are frequently very sensitive to damage in this zone because of intracellular freezing and procedures have been designed to permit cells to dehydrate fully at a temperature 10° or so above this zone before quenching in liquid nitrogen (*29*). These phenomena could have no relevance to a pure enzyme solution, so we must presume the zone relationships are purely coincidental. However, we do know that this zone is not unique for catalase in phosphate-buffered saline. We have done some preliminary studies with another tetrameric metalloenzyme which can only be purified and stored in strong salt solutions so the problem is not trivial. This enzyme, adenylate deaminase

Table IV. Recovery of Purified

EXP	*mg P/mL*	*KCl (m*M*)*	*NaCl (m*M*)*	*KPO_4 (m*M*)*	*MEt (m*M*)*
A	1	600	—	40	14
	1	600	—	40	1
B	0.5	600	—	40	0.5
	0.5	300	300	40	0.5
C	0.33	600	—	40	0.33
	0.33	200	400	40	0.33
D	0.1	600	—	40	0.1
	0.1	60	600	40	0.1

[a] After frozen storage at several temperatures in various media. All samples were quenched in alcohol at −72° before transfer to alcohol at the stated storage temperatures, and all were thawed rapidly in a 37° water bath for assay. All samples in a given experiment were prepared, frozen, stored, thawed, and assayed simultanously. ppt = precipitated on thawing. P = protein. MEt = mercaptoethanol.

(purified from human muscle), requires 0.6*M* salt levels, normally buffered with phosphate. Relying on our knowledge gained from the catalase studies, we used an all potassium system so that we could freeze safely at any temperature. We froze the samples with great confidence and destroyed them completely. As shown in Table IV this was due to the presence of 14m*M* mercaptoethanol which while beneficial at 4°, denatured the enzyme during freezing, so that thawed samples were not only inactive but had extensive precipitates. On dropping the thiol level to 1m*M*, this problem was eliminated and the enzyme could be stored successfully at a level of 1 mg/mL. (Mercaptoethanol also inactivates catalase on freezing, but this is hardly surprising since it also causes activity loss at 4°). When the adenylate deaminase was diluted 2-fold to 10-fold in the same medium, it became progressively more labile to damage during freeze-storage, especially at −18°. Diluted in phosphate buffered NaCl, rather than KCl, damage was most severe at −36°. As with catalase, then, lability increased with dilution; a low temperature zone of damage was present if the medium contained more sodium than potassium; and damage was much less at −70° than at higher temperatures.

Human Muscle Adenylate Deaminase[a]

	Activity Recovered after Storage (%)					
Days	*−18°*		*−36°*		*−72°*	
1	3 (ppt)		4 (ppt)		27 (ppt)	
1	103		103		93	
19	85		98		97	
80	—		—		90	
7	74		84		103	
7		48		38		82
3	60		94		94	
10	21		49		91	
3		55		58		82
10		38		8		67
3	30		54		80	
10	17		42		68	
3		22		0		51
10		0		0		40

Effects of Buffer Composition

The damage zone at −35° to −40° is more than 12° below the eutectic temperature for phosphate-buffered saline. Since it is operative for two quite different enzymes, the question arises as to whether it is unique to this medium, the one most widely used in biology and cryobiology. In Table V we compare the loss in catalase activity after one day storage at −40° and at −80° in 1% saline containing a variety of buffers at 4, 20, and 100m*M* levels. In brief, there is little enzyme damage in any system held at −80°, and there is little damage in tris, triethanolamine, citrate, and bicarbonate buffers at −40°. Imidazole and phosphate produced severe damage at −40° but were quite distinctive in behavior. Imidazole produced the same degree of damage in the absence of salt despite the fact that saturated solutions of this buffer were innocuous in the absence of freezing. In the case of phosphate, the presence of saline was essential for damage and the cation added was crucial. Damage was much reduced at high concentrations of added potassium phosphate but not at all with added sodium phosphate. Finally, note that potassium phosphate in 1% KCl results in negligible damage at −40° at all concentrations. Thus the low temperature damage zone appears only with certain buffers and in the case of phosphate-buffered saline is clearly dependent also on the K/Na ratio which according to phase diagram studies determines the degree of acidity produced during freezing.

This relationship is shown in more detail in Figure 10 where catalase damage after one day storage at −40° is shown as a function of the phosphate level added to 1% saline. Less than 1m*M* phosphate is sufficient to give marked damage and 3–4m*M* is sufficient to produce maximal

Table V. Effect of Various Buffers on Catalase Inactivation[a, b]

Buffer (pH 7.0) (in 170mM NaCl)	*Storage Loss After 1 Day at −40° (or −80°) (%)*					
	4mM		*20mM*		*100mM*	
Tris	14.7	(0.5)	17.8	(0.9)	10.8	(3.6)
Triethanolamine	3.6	(−0.1)	6.8	(3.7)	7.5	(7.6)
Na citrate	1.0	(−0.7)	−1.3	(−3.9)	21.8	(15.2)
Na bicarbonate	5.3	(17.2)	15.3	(−2.8)	26.3	(12.0)
Imidazole	65.4	(−0.3)	73.3	(−0.6)	70.4	(−1.0)
Na phosphate	81.0	(8.0)	88.0	(6.4)	90.0	(5.1)
K phosphate	81.0	(6.5)	86.0	(4.2)	23.0	(7.8)
K phosphate (in 170 m*M* KCl)	3.1	(10.4)	5.3	(9.2)	8.8	(6.7)

[a] In 1% saline.
[b] After one day storage at −40° and −80°. Tubes were seeded and quenched in liquid nitrogen before transfer to alcohol baths at the stated temperatures, and were thawed in a 37° water bath. Mean values are shown for 2 tubes/group.

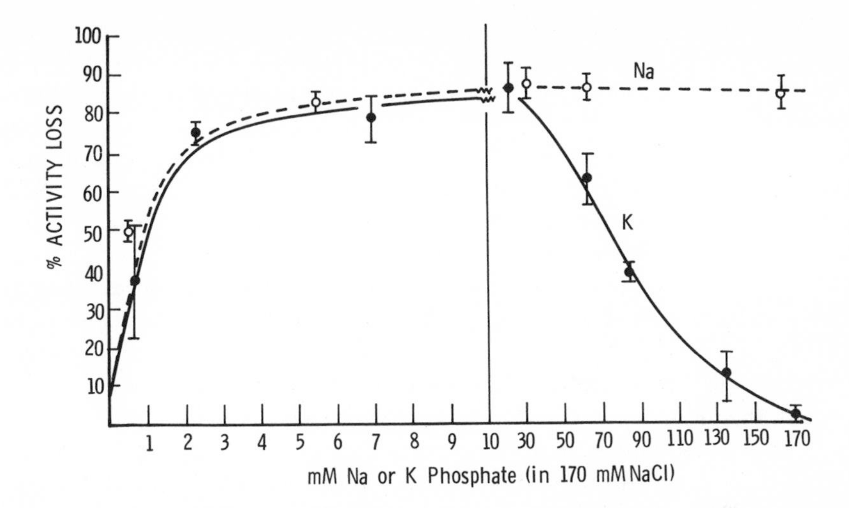

*Figure 10. Inactivation of catalase after one day storage at −40° in 170m*M *NaCl containing various additions of sodium or potassium phosphate (pH 7.0 at 25°). Experimental procedure was the same as described in Figure 9.*

damage which remains unchanged at all higher levels if sodium phosphate is added. However, additions of potassium phosphate above 30m*M* result in declining damage as the K/Na weight ratio approaches 1 and damage disappears altogether when the ratio reaches 3. This figure, therefore, taken together with phase diagrams for the phosphate-saline system, argue strongly that acidification is the factor responsible for damage. The only peculiarity is that the damage is occurring at a temperature far below the point of total solidification, according to the same phase diagrams. Since damage is progressive with time, we presume that there must be a liquidus, and hence that despite prior seeding and quenching, the freezing and thawing are not occurring under equilibrium conditions. It is well recognized that metastable states are quite common in frozen systems and may persist for many hours or even days, despite constant stirring (*3*). During the freezing of enzymes and cells where stirring is normally omitted, nonequilibrium behavior may be much more pronounced and prolonged. There has been increasing recognition of this factor recently and supplemented phase diagrams have been developed to incorporate such behavior (*30, 31*).

Acidity During Freezing and Thawing

The above considerations led us to explore directly the acidity in the frozen state with the use of water soluble pH indicators. A similar approach has recently been used by Orii and Morita to investigate buffer pH changes during freezing but not during thawing, using low-temperature spectrophotometry involving complex instrumentation (*32*). They found that most indicators tested were practically free from extrinsic color changes that would interfere with pH interpretations and that many buffers changed pH more than 3 units on freezing. We chose to use electronic flash photography of our sample tubes together with appropriate color standards, after which the developed transparencies could be used at leisure for comparison with the original tubes, visual evaluation, and spectrophotometry. We first employed Fisher Universal Indicator evaporated to its aqueous layer, and diluted 12-fold into the medium to give indicator levels of 6–80μM. Standards were prepared using pure K_2HPO_4 and NaH_2PO_4 which give essentially constant pH at all temperatures through the eutectic. Alternatively, standards were prepared by titrating the indicator in distilled water to selected pH values. With either set of standards, the color in the frozen state corresponded quite closely to that seen at 25° through frosted glass. With the basic validity of this approach thus established, the universal indicator permitted us to evaluate the pH range 4–9 with an estimated precision of 1 unit visually and 0.5 unit by spectrophotometric absorbance ratios at 550/675 nm.

With this procedure we noted that phosphate-buffered saline solutions remained neutral after seeding and after quenching to −80°, but became quite acid on subsequent storage at −40° whereas simple saline solutions remained neutral throughout. Four specific points were evaluated by absorbance ratios for the curve shown in Figure 10. At the two extreme points (saline alone, and saline plus 170m*M* potassium phosphate) where there was no damage, the pH remained 7.0 at −40°. In the presence of 1m*M* or of 90 m*M* potassium phosphate in saline, each of which gave 40%–50% damage, the pH reached 4.0–4.5 after one day storage.

Since these values are at the end of the acid range of the universal indicator, we chose Congo red at 35 μg/mL and a detection range of pH 3.5–5.5 for further studies. Figure 11 shows the striking color change at room temperature from red to blue as the pH of an aqueous solution is titrated from 5.0 to 3.0. Below that are seen the same colors when the tubes are wrapped in frosted tape to mimic the frozen state. Below that are shown the corresponding colors of the same solutions frozen and stored at −36°. The color match is quite satisfactory for the corresponding tubes at 25° and −36° and visual comparison can be used to estimate the pH in the frozen state. Figure 12 shows the color of this indicator in phosphate buffered saline stored one day at −80° and at −36°. Only at the higher storage temperature has the pH dropped below 4.0. If the system is totally solidified at −80°, then a liquidus must have reappeared at −36° in order for the indicator to have altered its ionization and reflect the increasing acidification. Moreover, this level of acidity would readily explain the denaturation of catalase and other enzymes on freezing.

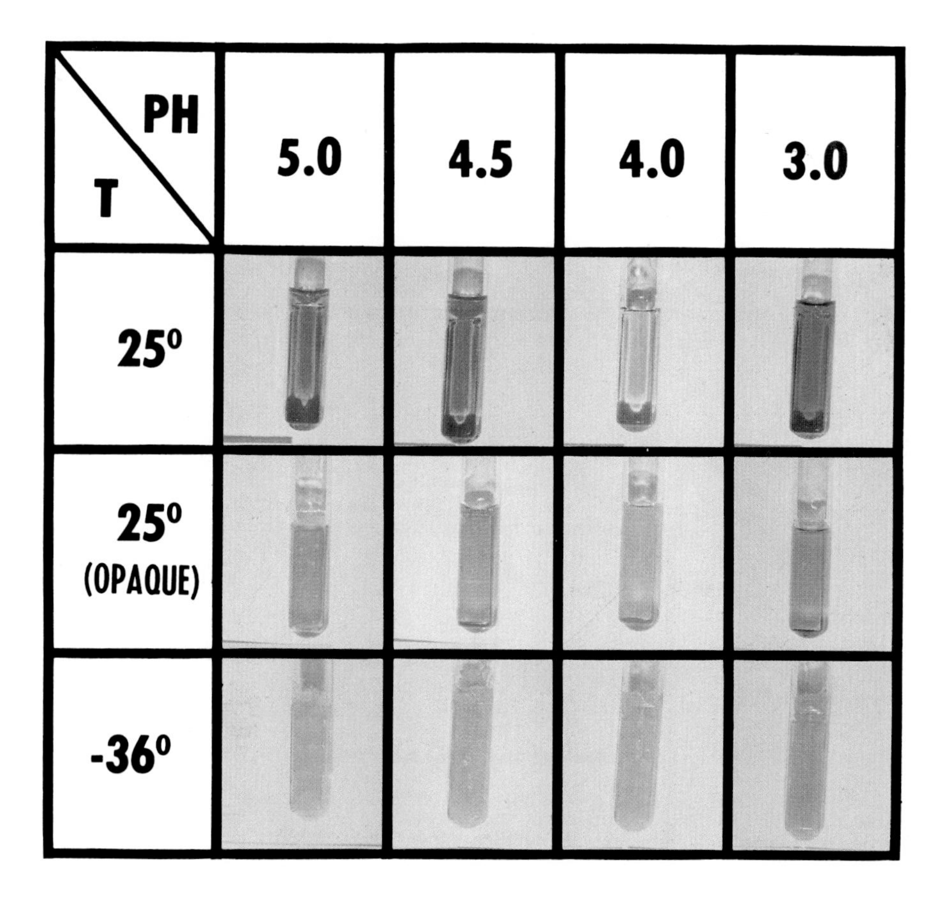

Figure 11. Appearance of 35 μg Congo red/mL distilled water titrated to various pH levels. Top row: Transparent tubes at 25°. Middle row: The same tubes wrapped in frosted tape to mimic the frozen state. Bottom row: Transparent tubes frozen at −36° for one day.

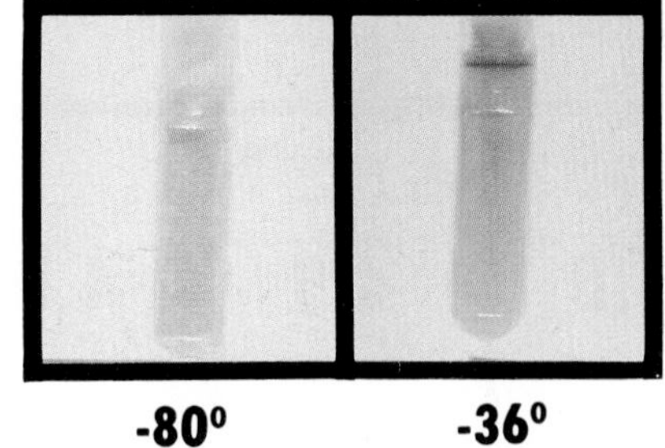

Figure 12. Appearance of 35 μg Congo red/mL 170mM NaCl + 10mM KPO_4 (pH 7.0 at 25°) stored one day at −80° and at −36° (after seeding and quenching at −80°).

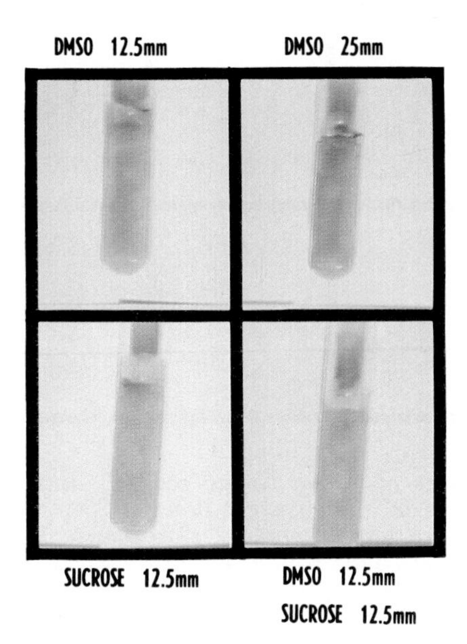

Figure 13. Effect of DMSO and sucrose, separately and together, on the appearance of 35 μg Congo red/mL 170mM NaCl + 10mM KPO_4 (pH 7.0 at 25°) stored one day at −36° (after seeding and quenching at −80°).

We can use this procedure to reevaluate the interaction of sucrose and DMSO in phosphate-buffered saline, recalling that DMSO exaggerated the damage to catalase at −36° while sucrose prevented it and reversed the DMSO effect. We now ask whether this may be due to pH effects. Figure 13 shows a corresponding experiment in which catalase has been replaced by Congo red. With 12.5 or with 25m*M* DMSO the pH is below 4.0 whereas with 12.5m*M* sucrose it is above 5.0. This was also true with higher levels of sucrose. In the mixture containing 12.5m*M* of each agent, the pH is again above 5. While we cannot say from these experiments that the DMSO lowers the pH as compared to phosphate-buffered saline alone, it is clear that the protection and reversal produced by sucrose is reasonably explained as a consequence of its decreasing the acidity at −36°.

From the standpoint of these studies we can differentiate three classes of cryoprotectants: the standard water-miscible solvents which produce a low-temperature-zone damage at low doses; macromolecular agents which do not but are ineffective in the presence of the solvents; and oligosaccharides which are protective in this zone at low concentrations and reverse the damage produced by standard agents. It is pertinent to note here that during studies on freezing of invertase which was labile at pH $<$ 4.8, Tong and Pincock (*8*) found that 25m*M* DMSO exaggerated the freezing damage in acid solutions while higher levels of DMSO decreased it.

Now the importance of acidity in damage to catalase in phosphate-buffered saline is not a surprise, except for the temperature at which it occurred and its reversal by sucrose. Because of these findings, however, we returned to the universal indicator in 10m*M* neutral potassium phosphate buffer alone in which we had begun these studies. As shown in Table VI, this buffer remained neutral on seeding and on quenching to

Table VI. Estimated pH Values during Freezing and Thawing [a]

		Indicator pH
A. Seeded, cooled at 0.7°/min to	−20°	6.1
B. Seeded, quenched in alcohol at	−80°	7.0
then equilibrated at	−20°	7.2
then warmed at 0.7°/min to	−5°	4.0
	−1°	5.2
	0°	6.3
	+1°	6.6
	+5°	7.0

[a] 10m*M* neutral potassium phosphate buffer containing a universal indicator. Values were calculated for absorbance ratios at 550 nm and 675 nm measured on color transparencies as described in the text. During warming, the solution remained partially frozen until +5°.

−80°. On slow warming the sample from −20°, the indicator remained neutral until the temperature reached about −5°, at which point it shifted to a pH of about 4.0, then gradually returned to neutral on thawing. This unexpected result indicated that acidification was also important in the neutral phosphate buffer system.

Interaction of Freezing and Thawing

Despite prior seeding, freezing and thawing do not usually occur under equilibrium conditions. In this event, the degree of disequilibrium may reflect the interaction of the two parts of the freeze–thaw cycle; the sample, so to speak, remembers its history. Table VII illustrates this for the thawing of fast-frozen versus slowly-frozen catalase samples. At any selected temperature in the thawing region, the fast frozen sample is more damaged. The damage is most severe at −1.5°, the same temperature at which the samples were initially seeded and held for the same time with negligible damage. Finally, at −8° damage was minimal and self-limited so that the slowly and fast frozen samples eventually sustained equal damage. We conclude that damage due to fast freezing occurs during thawing not during freezing.

Table VII. Activity Lost in Solutions of Catalase[a]

		Activity Lost (%)	
		Fast Frozen	*Slowly Frozen*
Held 20 min	−8°	8.5 ± 1.0	1.6 ± 2.7
Held 3 hrs	−8°	10.7 ± 3.5	12.8 ± 2.2
Held 20 min	−5°	28.8 ± 2.6	16.1 ± 3.1
Held 20 min	−1.5°	43.2 ± 1.3	27.1 ± 1.2

[a] 1.7 μg/mL 10m*M* potassium phosphate, pH 7.0 held in various thawing zones after slow or rapid freezing. All samples were seeded and held 20–30 mins at −1.5°. Four samples were then fast warmed in a 37° water bath and served as controls (these had 5%–10% activity loss compared to stock tubes). Sixteen tubes were then quenched and stored in ethanol at −70° (fast frozen, 90°/min). Sixteen tubes were cooled at 0.5°/min to −24°, then transferred to ethanol at −70° (slowly frozen). Four tubes from each group were then transferred to an alcohol bath at the temperature noted, after which all were fast thawed at 37°. Mean ± SE are shown for each group of four.

Conclusions

Now can we present an elementary framework on which to hang these isolated observations? We will present the simplest possible interpretation we can imagine, in our best Socratic manner. In dilute neutral solutions of potassium phosphate buffer the main intrinsic factor in freeze–thaw damage to catalase is the enzyme concentration while the main extrinsic factor is the warming rate. What are the explanations for these factors? The increasing lability to freezing damage on enzyme dilution parallels the inactivation behavior of catalase, and of most other enzymes in solution, and we invoke the usual explanation: higher concentrations provide increasing macromolecular interactions which retard inactivation produced by solvent, solute, pH, and interfacial effects. Slow warming is damaging because thawing occurs under nonequilibrium conditions, accompanied by acidification of the medium. The slower the warming rate, the longer the enzyme is exposed to the acid and the greater the damage. Note that we are presuming here that freeze–thaw damage occurs only while the enzyme is in solution and not while it is precipitated.

Well then, why is slow freezing not equally damaging? First, keep in mind that slow cooling is not equivalent to slow freezing. In the absence of prior seeding, slow cooling produces supercooling followed by spontaneous nucleation and rapid solidification so that the enzyme is but briefly exposed to any acid generated. On slow cooling after seeding, the system freezes close to equilibrium conditions and significant acidification does not occur. Another factor that may be of equal importance is the concentration of the enzyme during the two phases of the cycle. On slow freezing the enzyme is concentrated, and thus protected, as water is initially frozen out selectively from the solution. During thawing, however, the liquidus initially contains no enzyme; the enzyme redissolves gradually, presenting a dilute solution that is more readily inactivated. This combination of a dilute solution of enzyme in an acidifying medium explains why thawing damage does not begin until $-7°$ to $-8°$, although a liquidus should be present at $-17°$.

Well, that would explain why freezing itself is relatively innocuous, but why should rapid freezing produce increased damage? Two factors may explain this observation. First, the greater the freezing rate the greater the disequilibrium and acidification that will occur on subsequent thawing. Second, the greater the freezing rate the more likely is the enzyme to be trapped in multiple, small precipitant packets resulting in earlier resolvation upon thawing. Either or both of these factors result in greater thawing damage subsequent to rapid freezing.

Now what happens on adding NaCl to the medium? Under these conditions we are contributing large increments of concentrating salt as well as profound acidification over a much wider temperature range, during both the freezing and thawing phases of the cycle. Consequently, slow-cooling rates now produce marked damage during the freezing phase as the enzyme is exposed for longer periods before precipitating. This portion of damage becomes less marked as the freezing rate is increased but again the increasing rate promotes disequilibrium thawing with greater acidity and/or earlier resolvation of the enzyme, resulting in more extensive damage during the thawing phase.

Does it not follow then that we are explaining an optimum cooling rate recovery curve by the operation of a single damaging agent? In effect, yes: concentrating acid and/or salt are producing the damage at both extremes of the cooling rate curve. At very slow-cooling rates the damage is occurring primarily on freezing while at very rapid-cooling rates the damage is occurring primarily on thawing. The appearance of an optimum is due to the influence of the prior freezing rate on the composition of the medium and the rate of resolvation of the enzyme during subsequent thawing. The appearance of damage at temperatures far below the equilibrium eutectic indicates that the system, in fact, is operating far from equilibrium conditions, at least when slow thawing follows rapid freezing.

What then is the effect of adding a macromolecular cryoprotectant to this system? Such agents are effective at doses nearly stoichiometric with the catalase concentration and we interpret their effect as similar to that of increasing the enzyme concentration or to adding albumin to enzyme solutions, as is commonly done in biochemistry, to improve stability. The generic effect is to increase macromolecular interactions which retard solvent–solute inactivation.

But why should the usual low-molecular-weight cryoprotectants increase damage at low doses while protecting at higher doses? Low doses might increase the disequilibrium and acidity or might increase the enzyme solubility in the low-temperature zone without significantly diluting the toxic solutes. At higher doses the agents would increase the water content of the liquidus sufficiently to dilute toxic solutes and acidity, and thereby decrease damage.

How then could oligosaccharides reverse the potentiation of damage noted above? One simple possibility is that oligosaccharides may form new insoluble species with acid phosphate resulting in a pH closer to neutral and hence decrease damage whether or not other agents are present.

Now undoubtedly this picture, complicated as it is, is too simple to be a full explanation for cryodamage to enzymes; nevertheless, it may serve pro tempore as a basis for future experiments. In any event we can list the following suggested rules for cryopreservation of enzymes and other macromolecules in the absence of cryoprotectants: (1) keep the enzyme as concentrated as possible; (2) keep the volume of the samples small and the surface area large so that rapid thawing is feasible; (3) keep the content of salt, buffer, and other additives as low as possible; avoid buffers which markedly alter pH during freezing; in particular, avoid phosphate-buffered saline; (4) since nonequilibrium states are likely to occur in any case, freeze the sample quickly by quenching it in alcohol at −80° or better, in liquid nitrogen; (5) store the sample at −70° or below whenever possible; and (6) thaw the sample rapidly by swirling in a warm bath until thawing is complete.

Finally, to end this communication with the most succinct possible summary of factors involved in freezing damage to enzymes, we would have to say, the message is: the medium.

Acknowledgment

This investigation was supported in part by American Cancer Society Grant No. PDT-66B under the auspices of the American Registry of Pathology, Inc. The opinions or assertions contained herein are the private views of the authors and are not to be construed as official or as reflecting the views of the Department of the Army or the Department of Defense.

Literature Cited

1. Fishbein, W. N.; Winkert, J. W. *Cryobiology* **1977**, *14*, 389–398.
2. Fishbein, W. N.; Griffin, J. L. *Cryobiology* **1976**, *13*, 542–556.
3. Van den Berg, L.; Rose, D. *Arch. Biochem. Biophys.* **1959**, *81*, 319–329.
4. Hanafusa, N. In "Cellular Injury and Resistance in Freezing Organisms"; Asahina, E., Ed.; *Hokkaido University: Sapporo, Japan,* 1967; pp. 33–50.
5. Soliman, F. S.; Van den Berg, L. *Cryobiology* **1971**, *8*, 73–78.
6. Shikama, K. *Sci. Rep. Tohoku Univ., Ser. 4* **1963**, *27*, 91–106.
7. Brandts, J. F.; Fu, J.; Nordin, J. H. In "The Frozen Cell"; Wolstenholme, G. E. W., O'Connor, M., Eds.; J & A Churchill: London, 1970; pp 189–212.
8. Tong, M.-M.; Pincock, R. E. *Biochemistry* **1969**, *8*, 908–913.
9. Chilson, O. P.; Costello, L. A.; Kaplan, N. O. *Fed. Proc., Fed. Am. Soc. Exp. Biol., S-15,* **1965**, *24*, S555–S565.
10. Greiff, D.; Kelly, R. T. *Cryobiology* **1966**, *2*, 335–341.
11. Curti, B.; Massey, V.; Zmudka, M. *J. Biol. Chem.* **1968**, *243*, 2306–2314.
12. Irias, J. J.; Olmstead, M. R.; Utter, M. F. *Biochemistry* **1969**, *8*, 5136–5148.
13. Kono, N.; Uyeda, K. *J. Biol. Chem.* **1973**, *248*, 8603–8609.
14. Holland, M. J.; Westhead, E. W. *Biochemistry* **1973**, *12*, 2270–2275.
15. Graves, D. J.; Sealock, R. W.; Wang, J. H. *Biochemistry* **1965**, *4*, 290–296.
16. Singh, S.; Wildman, S. G. *Plant Cell Physiol.* **1974**, *15*, 373–379.
17. Chollet, R.; Anderson, L. L. *Biochim. Biophys. Acta* **1977**, *482*, 228–240.
18. Jarabak, J.; Seeds, A. E.; Talalay, P. *Biochemistry* **1966**, *5*, 1269–1278.
19. Lovelock, J. E. *Brit. J. Haematol.* **1955**, *1*, 117–129.
20. Fishbein, W. N.; Winkert, J. W. *Cryobiology,* **1978**, *15*, 168–177.
21. Lauffer, M. A. In "Subunits in Biological Systems, Part A"; Timasheff, S. N., Fasman, G. D., Eds.; Marcel Dekker: New York, 1971; pp 149–199.
22. Timasheff, S . N.; Frigon, R. P.; Lee, J. C. *Fed. Proc., Fed. Am. Soc. Exp. Biol.* **1976**, *35*, 1886–1891.
23. Frigon, R. P.; Timasheff, S. N. *Biochemistry* **1975**, *14*, 4567–4573.
24. Mazur, P. *Cryobiology* **1977**, *14*, 251–272.

25. Van den Berg, L. *Arch. Biochem. Biophys.* **1959**, *84*, 305–315.
26. Whittam, J. H.; Rosano, H. L. *Cryobiology* **1973**, *10*, 240–243.
27. Steponkus, P. L.; Garber, M. P.; Myers, S. P.; Lineberger, D. *Cryobiology* **1977**, *14*, 303–321.
28. Fishbein, W. N. *Cryobiology* **1972**, 9, 325.
29. Farrant, J.; Walter, C. A.; Lee, H.; McGann, L. E. *Cryobiology* **1977**, *14*, 273–286.
30. Pozner, R. I.; Shepard, M. L.; Cocks, F. H. *J. Mater. Sci.* **1977**, *12*, 299–304.
31. MacKenzie, A. P. *Philos. Trans. R. Soc. London, Ser. B* **1977**, *278*, 167–189.
32. Orii, Y.; Morita, M. *J. Biochem., Tokyo* **1977**, *81*, 163–168.

RECEIVED June 16, 1978.

Anomalous Depression of the Freezing Temperature by Blood-Serum Proteins of Fishes

ROBERT E. FEENEY and DAVID T. OSUGA

Department of Food Science and Technology, University of California, Davis, CA 95616

YIN YEH

Department of Applied Science, University of California, Davis, CA 95616

Two types of antifreeze proteins that function noncolligatively appear to exist in blood serum from fishes of polar oceans. One type found in Antarctica and the Arctic is a glycoprotein of repeating units of a glycopeptide composed of the tripeptide, Ala–Ala–Thr, with all threonines glycosidically α-linked to the disaccharide, β-D-galactopyranosyl-(1→3)-N-acetyl-D-galactosamine. The second type, a nonglycoprotein, exists in some northern fish. Ala constitutes approximately two-thirds of both types of protein. Both lower the freezing temperatures without affecting the melting temperature. A structural model has been suggested in which these antifreeze molecules shield the ice embryo surface. Roles are postulated for the carbohydrates and the methyl side chains of the glycoprotein and for the hydrophilic residues and methyl groups of the nonglycoprotein.

Poikilothermic organisms of marine waters may be exposed to temperatures that are below the freezing temperatures of the body fluids of fish from nonsalt waters. In fact, the freezing temperatures of ocean waters in equilibrium with ice are usually around $-1.8°$ to $-1.9°C$. Achieving such a low freezing temperature by the usual mechanism of

0-8412-0484-5/79/33-180-083$6.25/0

colligative action would require that their fluids have a concentration of NaCl (or its equivalent in other substances) equal to that of ocean water, namely around 3.2%. Such a high concentration of colligatively acting substances is at least 50% higher than that which many tissues can tolerate osmotically.

Marine organisms attempt to survive in ice-laden marine waters by several different routes. Some organisms, lacking the more advanced physiological systems such as those found in teleosts, actually can be frozen and then revived upon thawing. Such is the case with some marine mussels and snails (*1*). Other organisms, including some teleosts, can remain supercooled for prolonged periods unless seeded with ice crystals (*2, 3*). Fish in this latter category often are cited as the ones that freeze when caught in deeper waters and hauled to the surface where ice exists. Another route for several fishes has been the development of substances that lower the freezing temperature noncolligatively and therefore do not cause osmotic difficulties. These have been termed antifreeze substances (*4, 5, 6*) and are the subject of this chapter.

Two different types of antifreeze substances have been found in the bloods of fishes. The first of these, a glycoprotein called antifreeze glycoprotein (AFGP) (*4, 5, 6, 7, 8*), is found in cold fishes from both the Antarctic and the Arctic polar areas. Most of the information currently available concerns this substance. The second substance, a nonglycoprotein which we will term antifreeze protein (AFP), has been found in fish of cold but nonpolar waters (*8*). Although AFP is not a glycoprotein, it has a very important structural similarity to AFGP which will be discussed below. The presence of AFGP or AFP in a fish serum can be easily detected by determining the temperature of freezing of the serum before and after dialysis (*5, 6, 8*). A freezing temperature less than −1.5°C before dialysis and a freezing temperature somewhere below 0°C (usually less than −0.2°C) after dialysis is essentially conclusive evidence for the presence of a high-molecular-weight antifreeze substance (Table I) (*9, 10*). Still further confirmation of its presence is a determination of the melting temperature of the undialyzed serum. When antifreeze is present, the melting point is higher than the freezing temperature (*11*). (The term antifreeze protein is used in lieu of the term "freezing-point depressing protein" previously used (*9*), and the temperature at which its solution freezes is termed its freezing temperature rather than its freezing point. This is because the temperature of freezing is depressed, but the temperature of melting is not (*11*). The physical definition of the freezing (or melting) point is the temperature at which solid and liquid phases are in equilibrium, i.e., freezing and melting temperatures are identical, and these terms are therefore incorrect for this system.)

Table I. Freezing Temperature of Sera before and after Dialysis[a]

Species	Location	Freezing Temperature (°C) Before Dialysis	After Dialysis
Homo sapiens (human)		−0.56	−0.01
Oncorhynchus tschawtscha (king salmon)	California coast	−0.84	−0.01
Oncorhynchus tschawtscha (king salmon)	California river	−0.58	−0.01
Salmo gairdnerii (rainbow trout)	California hatchery	−0.65	−0.01
Mallotus villosus (capelin)	Barents Sea, Arctic	−1.18	−0.01
Lycodes vahli	Barents Sea, Arctic	−0.98	0.00
Liparis montagui	Barents Sea, Arctic	−1.26	0.00
Boreogadus saida (polar cod)	Barents Sea, Arctic	−2.11	−0.63
Trematomus borchgrevinki	Ross Sea, Antarctica	−2.07	−0.56
Dissostichus mawsoni	Ross Sea, Antarctica	−1.99	−0.49

[a] Data from Komatsu (*9*) and Osuga and Feeney (*10*).

Certain details of these substances have been presented previously. A general description of the research programs and of the procedures used to obtain fish in Antarctica has been given (*12*). Recently, one article summarized the more theoretical concepts of the mode of action of the antifreeze proteins (*7*) while a second, more extensive article was directed at protein chemists (*8*). The latter article describes in detail the general chemistry and properties of the antifreeze substances, as well as presenting hypotheses and models for their modes of action.

In this chapter the authors have attempted to focus on the properties and functions of the antifreeze substances as well as to summarize that information which has become available since the previous articles were prepared.

Earlier Studies on Antifreeze Proteins. Scholander and co-workers (*2, 13, 14*) found that the blood sera of Arctic fish had lower freezing temperatures than did the sera of fish not adapted to the cold. They reported nonsalt substances in the serum which helped to lower the freezing temperature. These were found in the fraction soluble in trichloroacetic acid upon chemical fractionation of the serum.

The existence of an antifreeze-like substance in northern polar fish has been confirmed more recently by Scholander and Maggert (*15*), Hargens (*16*), and Raymond et al. (*17*), who found it in the saffron cod, *Eleginus gracilis.* An antifreeze substance also has been reported in the winter flounder (*Pseudopleuronectes americanus*) (*18, 19, 20*). Our laboratory recently has found antifreeze glycoproteins in the polar cod (*Boreogadus saida*) from the Barents Sea north of Russia (*10*).

The first definitive report on southern polar areas was given in 1969 by DeVries and Wohlschlag (*21*), who reported on the antifreeze protein in Antarctic fish. The first extensive chemical studies on the antifreeze proteins were with proteins obtained from Antarctic fish (*22, 23, 24, 25*). These were done under the auspices of, and with the facilities of, the U.S. National Science Foundation (*12*).

Arctic Studies. Arctic studies on fish have been done for centuries and intensive studies have been done during the past 50 years. This is because of the proximities of humans to these areas and because of the

Table II. Freezing Temperature and Salt Concentration of Barents Sea Fish Sera[a]

Fish sera	Freezing Temperature (°C) Before Dialysis	Freezing Temperature (°C) After Dialysis	Conductance (ohms^{-1} × 500)	NaCl[b] (g/L)	Freezing Temperature (°C)[c]
Boreogadus saida[d]					
Fish 1	−1.94	−0.780	540	13.8	−0.831
Fish 2	−2.37	−0.638	600	15.5	−0.933
Fish 3	−2.06	−0.687	570	14.6	−0.879
Mallotus villosus[d]					
Fish 1	−1.47	−0.020	430	11.0	−0.662
Fish 2	−0.984	−0.019	440	11.2	−0.674
Fish 3	−1.10	0.0	470	11.7	−0.704
Liparis montagui[e]	−0.976	−0.002	310	8.0	−0.482
Lycodes vahli[e]	−1.26	0.0	460	11.5	−0.689

[a] Data from Osuga and Feeney (*10*). Fish sera (0.2 mL) were checked for freezing temperature before and after dialysis immediately after the preparation of the sera. Conductance was measured by diluting 0.010 mL of sera to 5.0 mL with H_2O and measured by a dip-type conductivity cell attached to a conductivity bridge.

[b] NaCl concentrations were calculated by assuming that the conductance was only caused by the presence of NaCl in the sera.

[c] Freezing temperature calculated from the NaCl concentrations.

[d] The samples were taken from the same pooled sera on which the physiological fluid analyses were done (*see* Table III).

[e] Determination on the serum from individual fish.

Journal of Biological Chemistry

economic importance and food supply of these areas. Consequently, most of the numerous northern studies have concerned surveys of fish populations or biological investigations aimed at ensuring an adequate food supply.

The Arctic program of the authors' laboratories began in 1976 when two of us were guests for nearly a month aboard the Norwegian research vessel, the G. O. Sars. The blood of four species from the Barents Sea and Arctic Ocean were examined. From one of these—*B. saida,* obtained at 62°E and 78°N—AFGP was isolated and characterized (*10*). The bloods of fishes contain many small molecules which lower the bloods' freezing points, including amino acids and salts. However, as seen from Tables II and III, upon examination of samples of sera from two Arctic fishes, *B. saida* (containing AFGP) and *Mallotus villosus* (not containing an antifreeze substance), the only noteworthy difference affecting freezing temperature was a small increase in conductivity in the *B. saida* serum (*10*). This was attributable to salt while the much higher amount of D,L-*O*-phosphoserine was still too small to materially affect the freezing temperature.

Antarctic Studies. Possibly some of the most constant temperatures of ice-ocean waters exist in the Ross Sea of Antarctica, particularly on the western end of Ross Island along the edge of the Ross Ice Shelf in McMurdo Sound. The U.S. National Science Foundation has operated a biological research laboratory at McMurdo Base (a U.S. Navy installation) for nearly two decades, and nearby is the smaller New Zealand installation, Scott Base (*12*). The Antarctic program of the authors' laboratories began in 1964 under the auspices of the U.S. National Science Foundation at McMurdo Base on Ross Island (*12*). Research programs varying from two to five months in duration were conducted in each of seven different years in that vicinity.

There have been studies on at least one dozen species of fishes in McMurdo Sound. Those most studied by Antarctic investigators are *Trematomus borchgrevinki, T. bernacchii,* and *Dissostichus mawsoni.* The characteristics of the fish are discussed elsewhere (*8, 12*).

Antifreeze Glycoproteins

Occurrence in Antarctic Fish Serum. The AFGP of two Antarctic fish, *T. borchgrevinki* and *D. mawsoni,* have been studied in detail. The low-blood-serum freezing temperature of the *T. borchgrevinki* is attributed 70% to dialyzable solutes and 30% to nondialyzable solutes which is AFGP (Table IV) occurring in the serum at a total concentration of ~ 15 mg/mL. The AFGP from these two very different Antarctic fishes have very similar structures (*22, 23, 24, 25*). AFGP is a collective **name**

Table III. Physiological-Fluid Analysis on Sera

Amino Acids (Ninhydrin Reacting Constituents	*Mallotus villosus*[b]				
	1[c]	*2*[c]	*3*[c]	*Av*	*%*
DL-*O*-Phosphoserine	1.1	1.7	3.2	2.0	0.25
Taurine	113	127	164	135	17
Aspartic acid	19	22	9.1	17	2.2
Threonine	37	55	32	41	5.2
Serine	23	30	8.3	20	2.5
Glutamic acid	17	17	25	20	2.5
Glutamine	26	34	5.1	22	2.8
Proline	15	31	14	20	2.5
Glycine	79	97	73	83	10.6
Alanine	68	104	61	78	9.9
Citrulline	0.6	4.0	1.4	2.0	0.25
α-Amino-*n*-butyric acid	2.3	1.9	1.1	1.8	0.23
Valine	44	87	51	61	7.8
Methionine	1.7	5.5	1.0	2.7	0.34
Cystathionine	1.1	2.2	0.3	1.2	0.15
Isoleucine	18	35	19	24	3.1
Leucine	39	71	42	51	6.5
Tyrosine	12	20	11	14	1.8
Phenylalanine	12	18	11	14	1.8
β-Alanine	nd[d]	[e]	[e]	nd	nd
Tryptophan	1.8	2.5	1.2	1.8	0.23
Ethanolamine	10	12	10	11	1.4
Ammonia	88	84	116	96	12
Ornithine	3.1	6.1	4.5	4.6	0.58
Lysine	13	51	30	31	3.9
Histidine	9.4	12	14	12	1.5
1-Methylhistidine	3.1	1.7	2.0	2.3	0.29
Arginine	6.9	33	17	19	2.4
Total	664.1	965.6	727.2	787.4	100

[a] Data from Osuga and Feeney (*10*). Equal volumes of pooled serum (40 μL) and 6% sulfosalicyclic acid were mixed, centrifuged for 5 min at 8000 × g, and the entire supernatant analyzed on a Beckman amino-acid analyzer by the physiological-fluid system. The units used are μmol/100 mL serum.

[b] All sera samples contained approximately 1 μmol of urea per 100 mL of serum in addition to small amounts of numerous unidentified ninhydrin positive components.

[c] Pooled samples of sera from fish caught at one trawl catch. The approximate temperature and depth of the catch were: *M. villosus* (1) −1.5°C, 120 m, (2) 2.0°C, 30 m, (3) −1.3°C, 190 m; *B. saida* (1) −1.0°C, 160 m, (2) −1.0°C, 160 m, (3) 1.1°C, 150 m.

for a family of at least eight closely related glycoproteins. They were named AFGP-1 to AFGP-8 according to their relative migrations on gel electrophoresis in borate buffer (Figure 1). AFGP-1 to AFGP-5 (the larger ones) have antifreeze activity when tested alone. AFGP-6 to AFGP-8 initially tested negatively for antifreeze activity, a result which

from *Mallotus villosus* and *Boreogadus saida*[a]

Boreogadus saida[b]				
1[c]	*2*[c]	*3*[c]	*Av*	%
61	96	73	77	6.0
139	178	227	181	14
36	38	42	39	3.0
47	56	56	53	4.1
60	52	69	60	4.7
36	45	42	41	3.2
33	30	29	31	2.4
23	23	27	24	1.9
114	87	104	102	7.9
129	136	160	142	11
1.9	7.0	2.5	3.8	0.30
5.3	1.0	2.6	3.0	0.23
37	58	52	49	3.8
8.8	17	20	15	1.2
2.2	4.7	2.2	3.0	0.23
24	32	30	29	2.3
70	103	93	89	6.9
19	31	25	25	2.0
18	27	26	24	1.9
18	23	[e]	21	1.6
2.3	3.0	3.3	2.9	0.23
7.8	3.4	11	7.4	0.58
105	69	69	81	6.3
0.5	1.8	0.9	1.1	0.09
102	119	139	120	9.3
11	17	14	14	1.1
1.8	1.5	1.3	1.5	0.12
49	50	57	52	4.1
1161.6	1309.4	1377.8	1291.7	100

[d] Not determined (nd).
[e] β-Alanine had a constant factor for some 15 times less than phenylalanine, thus resulting in a very low sensitivity of β-alanine. An estimated amount of β-alanine was ~ 5 μmol/100 mL of sera.

Journal of Biological Chemistry

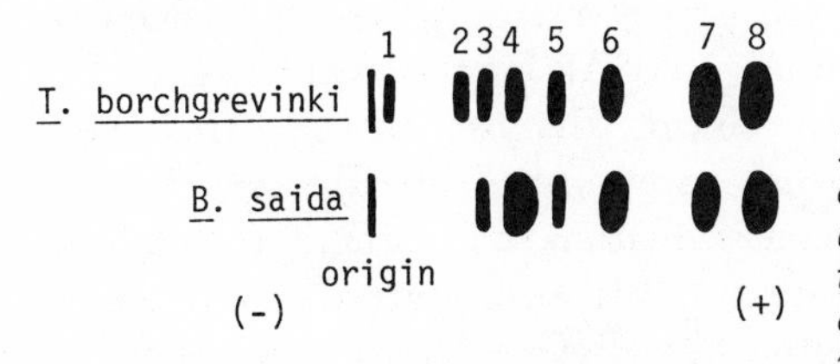

Figure 1. Reproduction of slab-gel electrophoretic pattern representing all of the major antifreeze glycoproteins present in the T. borchgrevinki *and* B. saida. *The numbers on top refer to the components present in* T. borchgrevinki *(10).*

Table IV. Properties of Antifreeze

Glycoprotein Component[b]	*Mol Wt*[c]	*Number of Triglycopeptides*[d]
1	32,000	50
2	29,000	45
3	21,500	35
3,4[e]		
4	17,500	28
5	10,500	17
	11,000 ± 2,000[f]	
6[g]	7800	12
7[g]	4000	6
8[g]	2600	4
	2646[i]	

[a] Summarized from Feeney (*6*) and Feeney and Yeh (*8*).
[b] Values for Glycoproteins 2, 3, and 4 are probably correct within ±5%; those for 6, 7, and 8 are slightly more accurate: data of Glycoprotein 1 is less accurate than the others. The inaccuracies are possibly caused by difficulties in the determinations of glycoproteins, and all preparations contained small and varying amounts of impurities.
[c] Data determined by sedimentation equilibrium unless otherwise noted.

puzzled investigators (*6*). However, recent studies show that AFGP-7 or AFGP-8 have extensive antifreeze activity as mixtures with AFGP-1 to AFGP-5 (*26*). (In this chapter we will refer to "active" AFGP for mixtures of one or more of Components 1–5, or to the actual components present, e.g., AFGP-4 or AFGP-1 to AFGP-5.)

Structure and Physical Properties of Antarctic AFGP. There are only two amino acids, alanine and threonine, in AFGP-1 to AFGP-5. The fundamental structure is that of a glycotripeptide of two alanines followed by a threonine with the sugars glycosidically linked to the threonines (Figure 2). The disaccharide is galactosyl-*N*-acetylgalactosamine with a β, 1→3 internal linkage (*25, 28*). This fundamental subunit is repeated approximately 17–50 times in AFGP-1 to AFGP-5 molecules. One or two alanines are added at the NH_2-terminal end of each polymer.

AFGP-1 to AFGP-5 is comprised of several polymers of molecular weights greater than 10,000 (Table IV). AFGP Components 3, 4, and 5 also will not pass through 3/32-in. Visking dialysis tubing. However, some heterogeneity of the individual components usually was found, perhaps because of some slight contamination with other recognized AFGP components or with small amounts of unrecognized AFGP polymers with differing polymeric lengths. Small amounts of a component provisionally named AFGP-5′, for example, have been found (*29*).

Glycoproteins of *Trematomus borchgrevinki*[a]

Antifreeze Activity	*Other Properties*
strong	
strong	
strong	
strong	viscosity, [n] (17°C) = 20 cc/g; no α-helix
strong	diffusion constant (20°C) = 5.64×10^{-7} cm^2 sec^{-1}; radius of gyration, R_e = 37 Å
strong	
weak	
potentiates[h]	
potentiates[h]	diffusion constant (20°C) = 12.1×10^{-7} cm^2 sec^{-1}; radius of gyration, R_e = 18.8 Å; viscosity, [n] (17°C) = 5 cc/g

[d] Values for number of triglycopeptides are calculated from molecular weights. Complete weight of glycoprotein also includes additional two NH_2-terminal alanines.
[e] Determination on mixtures of 3 and 4 (*22*).
[f] Determination by membrane osmometry (*22*).
[g] Glycopeptides 6, 7, and 8 also contain Pro.
[h] AFGP-7 and AFGP-8 have very weak activity when tested alone but show extensive antifreeze activity in mixtures with AFGP-1 to AFGP-5 under certain conditions.
[i] Determination by amino-acid sequence (*28*).

The small AFGP components, 6, 7, and 8, all had some prolines following threonines, and existed, at least primarily, as easily separable species. Sequence heterogeneity has been found with regard to the prolines (*28*) (Figure 3).

AFGP components are in an expanded state as based on data from intrinsic viscosity, circular dichroism spectra, and diffusion by ultracentrifugation or quasi-elastic light scattering. Examination of Raman spectra

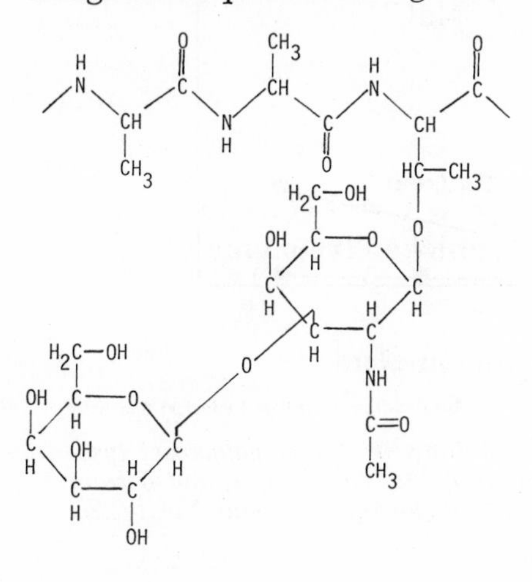

Figure 2. Polymer unit of the active antifreeze glycoprotein (27).

T. BORCHGREVINKI:

(1) NH_2–ALA–ALA–THR–ALA–ALA–THR–PRO–ALA–THR–ALA–ALA–THR–PRO–ALA–COOH

(2) NH_2–ALA–ALA–THR–ALA–ALA–THR–ALA–ALA–THR–PRO–ALA–THR–PRO–ALA–COOH

(3) NH_2–ALA–ALA–THR–ALA–ALA–THR–ALA–ALA–THR–ALA–ALA–THR–PRO–ALA–COOH

1 2 3 4 5 6 7 8 9 10 11 12 13 14

B. SAIDA:

NH_2–ALA–ALA–THR–PRO–ALA–THR–ALA–ALA–THR–PRO–ALA–THR–ALA–ALA–COOH

1 2 3 4 5 6 7 8 9 10 11 12 13 14

Journal of Biological Chemistry

Figure 3. Amino-acid sequences of Glycopeptide 8 isolated from the blood sera of T. borchgrevinki *(28) and* B. saida *(10).* T. borchgrevinki *samples apparently exist as a mixture of glycopeptides as indicated in Sequences 1, 2, and 3 with a relative distribution of 7:2:1 respectively (28).* B. saida *consists of only one sequence (10). Positions 7, 10, and 13 of the* T. borchgrevinki *are underlined to indicate the varying positions of prolines (10).*

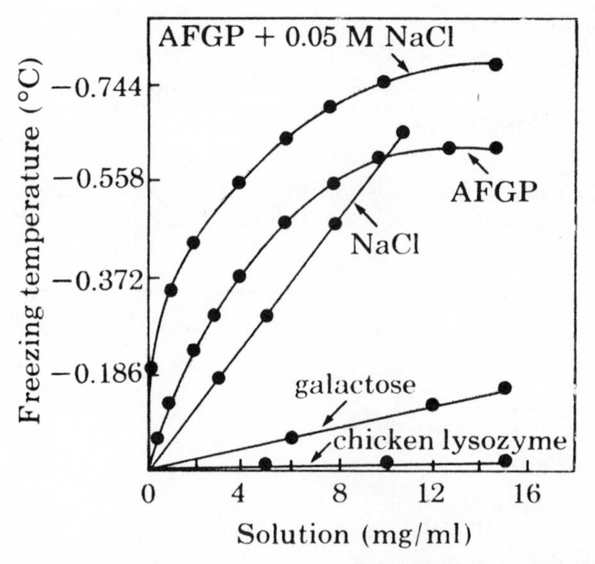

Adapted from Journal of Biological Chemistry

Figure 4. Freezing temperatures of solutions of NaCl, galactose, lysozyme, and a mixture of antifreeze glycoproteins (AFGP 3, 4, and 5 from T. borchgrevinki *(22, 26)), and a mixture of glycoprotein and NaCl (22).*

suggested some structures with α-helical and considerable β-conformation (*30*). Franks and Morris (*31*) recently suggested that the seemingly disordered conformation as evidenced by circular dichroism data may be explained also by a fixed conformation with a comparable degree of dissymmetry to the β-conformation but with substantially different molecular geometry. They have suggested that perhaps the high carbohydrate content of AFGP can force such a conformation where the hydrophobic side of the carbohydrates is associated with the peptide backbone while the hydrophilic side is exposed to the H_2O.

Preparations of AFGP lower the freezing temperature more than 200 times that which would be expected based on its molecular size (Figure 4). In fact, mixtures of these preparations at lower concentrations can be more active than an equal weight of NaCl.

Occurrence, Structure, and Properties of Arctic AFGP. Much less information has been available for the structures and properties of the Arctic antifreeze proteins as compared with those of the Antarctic ones. The glycoprotein of *E. gracilis* has a composition qualitatively similar to the AFGP from the Antarctic species, but definitive structures have not been reported (*19*). The recent studies on the AFGP of *B. saida* made detailed comparisons with the AFGP from the Antarctic fish *T. borchgrevinki* (*10*). The composition and sequence of the larger active fraction were similar, if not identical, in both species, although there were differences in the number of multiple molecular forms present (Figure 1). Both the larger and smaller glycoproteins contained alanine, threonine, and disaccharide. Although the distributions of different lengths of the peptides varied between the species, the larger antifreeze glycoproteins from both polar regions had the identical fundamental structure consisting of repeating units of the glycopeptide Ala–Ala–Thr, with all the threonines glycosidically α-linked to a β, 1→3 galactosyl-*N*-acetylgalactosamine. The smaller AFGP-6 to AFGP-8 also contained prolines following some of the threonines, but the positions were different in the Arctic and Antarctic materials and there were differences in the amounts of proline present.

Cooperative Functioning in Mixtures of Larger and Smaller AFGP. In a recent study (*26*), small amounts of the smaller AFGP-7 or AFGP-8 were added to even less amounts of AFGP-1 to AFGP-5. With this addition, the mixture gave rise to a dramatic enhancement of antifreezing capacity, which we have called potentiation and have interpreted as a cooperative function. In the presence of 10 mg of AFGP-8, the potentiation of the lowering of the freezing temperature with 2 mg of active AFGP was approximately fourfold, whereas with 25 mg of AFGP-8 the potentiation was more than eightfold. With still higher amounts of AFGP-8, even greater effects were observed. However, at levels of 6–8 mg/mL of active AFGP, the potentiating action of even 25 mg of AFGP-8 was small and the freezing curves began to plateau and merge.

Further studies have shown that potentiation occurs only in supercooled solutions (*32*). In solutions with negligible supercooling, AFGP-8 has weak activity by itself, but in supercooled solutions its activity is expressed only in the presence of the larger, active AFGP-1 to AFGP-5.

Antifreeze Proteins (Nonglycoprotein)

Occurrence of Nonglycoprotein Antifreeze Proteins (AFP). The Sculpin (*M. verrucosus*) contained an antifreeze substance that had amino acids other than alanine and threonine and no carbohydrate, but which was similar to the antifreeze glycoprotein in that alanine comprised approximately two-thirds of its total amino acids (*17*). In addition to alanine, appreciable amounts of aspartic acid or asparagine, glutamic acid or glutamine, leucine, lysine, and threonine, as well as other amino acids, were reported.

The laboratories of DeVries (*17*) and Hew (*20, 33*) reported a similar antifreeze protein from the northern flounder (*P. americanus*), which resides in the northern Atlantic coastal waters of America and Canada. Both confirmed that it is not a glycoprotein and that it has a high alanine content approximately similar to the antifreeze protein from *M. verrucosus.*

Three active components of flounder AFP with molecular weights estimated at 3000, 8000, and 12,000 by SDS gel electrophoresis were reported by Duman and DeVries (*18*). A. L. DeVries' (*18*) and C. L. Hew's laboratories (*20, 33*) have found that flounder AFP has almost the identical ratio of alanine to other amino acids as does AFGP, i.e., 2:1. The protein contains only eight amino acids with a reported composition in residues per 10,000 g of: Ala, 64.6; Arg, 2.0; Asp, 14.0; Glu, 1.7; Leu, 5.6; Lys, 2.9; Ser, 3.4; Thr, 11.2 (*18*). One reported sequence of the smaller component is (*34*):

10
NH_2–Asp–Thr–Ala–Ser–Asp–Ala–Ala–Ala–Ala–Ala–Ala–Leu–

15 20 25
Thr–Ala–Ala–Asx–Ala–Ala–Ala–Ala–Ala–Lys–Leu–Thr–Ala–

30 35
Asx–Asx–Ala–Ala–Ala–Ala–Ala–Ala–Ala–Thr–Ala–Ala–COOH

A partial sequence has been reported for the first N-terminal sequence of a main component (*20, 33*). This sequence is:

Asp–Thr–Ala–Ser–Asp–Ala–Ala–Ala–Ala–Ala–Ala–Leu–Thr–
Ala–Ala–Asn–Asx–Lys–Ala–Ala–Ala–Glu–Leu–Thr–Ala–Asp–
Asn–Lys

Preliminary experiments with Carboxypeptidase A digestion also indicate the presence of a cluster of alanines at the C-terminal.

Ananthanarayanan and Hew (*35*) have shown through circular dichroism measurements that flounder AFP possesses a large proportion ($\sim$ 85%) of an α-helical conformation at a low temperature ($-1°C$). The helical content decreases as the temperature rises. Viscosity data at $-1°C$ indicate an asymmetrical shape for the protein molecule compatible with its high helical content. The data for circular dichroism has recently been confirmed (*36*). No data is available concerning the assumption that this highly helical structure at $-1°C$ is related to its function.

Mixtures of AFP and AFGP-1 to AFGP-5 recently have shown additive increments in lowering the freezing temperature (*37*). Thus, they appeared to act independently of one another. However, as might be expected, mixtures of AFP and AFGP-8 did not exhibit potentiation.

The levels of AFP in the serum are the highest in January and February when the water is the coldest and low or absent in the summer (*33, 38*). Its seasonal synthesis and clearance from the serum also are affected by hypophysectomy (Figure 5) (*39*). Hew and Yip have pre-

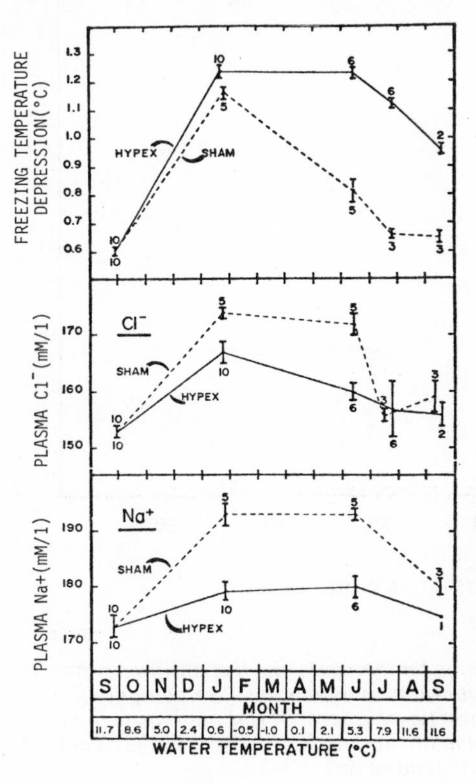

Canadian Journal of Zoology

Figure 5. Seasonal changes in plasma-freezing-temperature depression and Na^+ and Cl^- concentrations of hypophysectomized (HYPEX) and sham-operated winter flounder. Values are plotted as means ±1 standard error. Numerals at each point represent the number of fish sampled. The initial values for September (S) are pre-operated concentrations. Modified from Fletcher et al (39).

Table V. Properties of the Synthetic Polypeptide[a]

Sample No.	Composition (mol 1%)[b] Alanine	Aspartic Acid	Mol Wt[c]	α-Helical Content[d]
SAF-I[e]	65.5	34.5	45,000-70,000	15
SAF-II[f]	64.8	35.2	50,000-70,000	20

[a] From Ananthanarayanan and Hew (*40*).
[b] Determined on a Beckman Model 121M amino-acid analyzer after 24-48 hr of hydrolysis in 12*M* HCl at 110°C. The compositions of the polypeptides before deblocking the side-chain benzyl group were almost identical to those of the final products.
[c] Estimated using a Sephadex G-75 column calibrated with protein markers.
[d] Estimated on the basis of the circular dichroism data of Ananthanarayanan and Hew (*35*).
[e] Synthesized using HF-pyridine debenzylation method; SAF, synthetic antifreeze.
[f] Synthesized using the HBr debenzylation method.

Nature

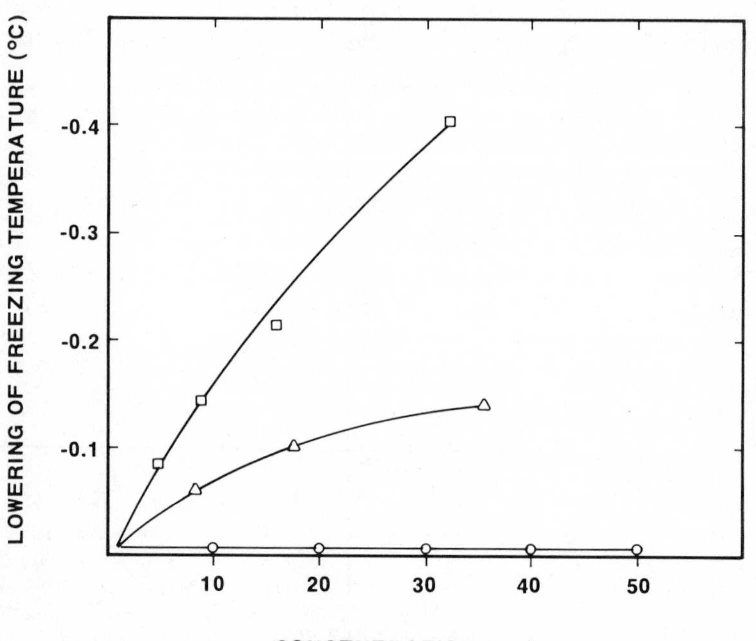

Nature

Figure 6. Lowering of the freezing temperature by the antifreeze protein of the flounder (□) and a synthetic polypeptide (△) composed of alanine and aspartic acid. The synthetic polypeptide is a random polymer of alanine and aspartic acid in the molar ratio of 2:1 (sample SAF-1, Table V) (40). (○, albumin)

sented data indicating that AFP is synthesized ribosomally (*20*). When a 6–10 S, poly A-rich RNA from the liver of the flounder was injected into *Xenopus oöcyte,* the RNA preparation stimulated a fourfold incorporation of [^{3}H] alanine into the antifreeze protein fraction.

Ananthanarayanan and Hew (*40*) recently have reported synthesis of a polypeptide with structural similarities to the AFP from the flounder (Table V, Figure 6). The product is a random polymer consisting of alanine and aspartic acid in a molar ratio of 2:1. Although it is stated to have only approximately one-third the antifreeze activity of the naturally occurring flounder AFP, controlled syntheses of such synthetic polymers could be one of the most promising ways to investigate the structural requirements for antifreeze activity. This synthetic approach should be especially valuable for determining the roles of the hydrophobic and hydrophilic groups as found in AFP and AFGP.

Studies on the Mechanism of Functions of Antifreeze Glycoproteins

Essentially all the available studies attempting to answer the question of how these proteins function to lower the freezing temperature noncolligatively have been done with AFGP and not AFP. The reason for this simply has been that AFGP has been available in much larger quantities and for a much longer time than AFP has. Although these relationships now should change with the availability of synthetic AFP (*40*), we shall at this time restrict this section to studies with AFGP from the Antarctic fish *T. borchgrevinki.*

Size of the Peptide Chain. Treatment of AFGP-5 with subtilisin or fungal protease resulted in loss of the antifreeze activity (*23*). When this loss was plotted against the number of amino groups present and a line was extrapolated to zero activity, it was found that the antifreeze activity was lost when an average of 1.5 to 1.8 peptide bonds per 10,000 g of AFGP had been cleaved. Progressive degradations of AFGP Component 5 by the Edman procedure caused no significant losses of antifreeze activity for the first three or four degradations (*6, 27*). After six or seven degradations approximately 85% activity was retained. A single attempt to estimate the minimum size of the polypeptide chain needed for antifreeze by partial proteolysis and separations of large fragments indicated that chains shorter than two-thirds of the smaller active AFGP Component 5 were inactive. Thus, a minimum size with greater than approximately 10 glycotripeptides or 30 amino acids appeared necessary.

Proteolytic fragmentation of AFGP-8 destroyed the potentiation in mixtures with AFGP-1 to AFGP-5 (*26*). When AFGP-8 was treated with the enzyme elastase, its potentiating activity on active AFGP was destroyed completely. Similarly, when AFGP-1 to AFGP-5 was treated with elastase, it could not be potentiated by native AFGP-8. In one such experiment, the freezing temperatures of AFGP-1 to AFGP-5 (2 mg/mL), AFGP-8 (15 mg/mL), and the solution of both, were −0.033°C, −0.016°C, and −0.185°C, respectively. After the AFGP-8 was treated with elastase, its freezing temperature was −0.022°C, and the solution of both was −0.063°C. Therefore, the elastase-treated AFGP-8 had < 10% of its original potentiating activity.

Covalent Modification of Carbohydrate Chains. With the exception of one modification, all covalent changes of the carbohydrate side chains of AFGP-1 to AFGP-5 resulted in loss of activity (*4, 22, 23, 25, 41–44*). Those causing losses of activity included acetylation of > 30% of the hydroxyls, oxidative destructions by periodate, oxidation of the C-6 hydroxyls to the carboxyl group by bromine, and β-elimination of the disaccharides with alkali or acetonation. However, removal of the acetyl group from the acetylated AFGP caused complete recovery of activity. The one modification not affecting activity was the oxidation of the C-6 hydroxyls of the galactose to the aldehyde group by treatment with galactose oxidase. Formation of the bisulfite adduct to this AFGP polyaldehyde resulted in complete loss of activity which could be regained by reversibly removing the sulfite in acidic solution.

Other enzymatic changes involving the carbohydrate chains also have caused losses in activities. Additions of sialic acid to the galactoses of AFGP-1 to AFGP-5 reduced the activity to less than 10% (*41*). Subsequent removal of the sialic acid resulted in considerable regaining of the original activity (45–75%), depending upon the amounts of sialic acid removed. Removal of the disaccharides from AFGP-8 caused a loss of more than 90% of the potentiation observed in mixtures with AFGP-1 to AFGP-5.

Formation of Complexes with Borate. AFGP-1 to AFGP-5 activity was lost when 2 mol of borate were bound per disaccharide side chain and these losses and bindings were pH dependent (*43*). Both the activity and the amount of borate bound to the antifreeze glycoprotein were influenced strongly by the pH of the solution between 7.0 and 9.0. The number of moles of borate bound per disaccharide unit was only about 0.5 at pH 8.0 but approached 2 at pH 9.0. When approximately 20% of the antifreeze activity was lost, 0.5 mol of borate were bound per mole of disaccharide and almost 100% of the activity was lost when 2.0 mol of borate were bound. The effects of pH and borate on the conformation of the antifreeze glycoprotein were studied in an effort to see whether or not the inactivation was a result of a conformational change on the binding of borate. No significant changes were observed in the diffusion coefficients as determined by quasi-elastic light scattering or equilibrium centrifugation or in the s_{20} values.

Unfortunately, none of the above studies proved unequivocally that the carbohydrate side chains were or were not required for function. However, they indicated that losses of activity could be caused by positioning numerous negatively charged groups (e.g., carboxyls, borates, or sulfites) on the side-chain sugars.

The Physical Aspects of the Action of Antifreeze Glycoprotein

The antifreeze action of AFGP is different in many respects from the action of colligatively functioning substances or the effects found in supercooled solutions.

Freezing Temperatures. Freezing temperatures of aqueous solutions of Glycoproteins 3, 4, and 5 as a function of concentration are shown in Figure 4 (*22*). Only one curve is shown because each of the glycoproteins depressed the freezing temperature to the same extent. In other experiments the freezing-temperature curve in the presence of 0.05*M* NaCl paralleled that for solutions of AFGP alone, indicating that the salt does not interact with the glycoproteins in a manner which influences the antifreeze activity of the glycoproteins. When the antifreeze activity was measured in the presence of buffers over the pH range from 3 to 12, the same additive effect was noted. At low concentrations AFGP was as efficient as NaCl but at higher concentrations ($>$ 10 mg/mL) AFGP was much less efficient, and the plot curved into nearly a flat plateau.

The Hysteresis between Freezing and Melting. A detailed study of the differences in freezing and melting temperatures of AFGP solutions has proven an absolute hysteresis with different sharp-freezing temperatures and melting temperatures (Table VI) (*11*).

Possible Functions as an Inhibitor of Nucleation. The prevention of freezing by AFGP in the presence of ice crystals would indicate that AFGP must function also at a post-nucleation stage, but it has not been proven whether or not AFGP functions to prevent nucleation.

However, one set of attempts to prove that inhibition of nucleation was the basis of function by AFGP was unsuccessful. These data were obtained by differential thermal analytical experiments (*11*).

Distribution of AFGP between Ice and Liquid Phase. Duman and DeVries (*45*) found that the concentration of AFGP-1 to AFGP-5 in the liquid phase was unchanged upon partial freezing of the solution and concluded that the active glycoprotein is incorporated into the ice phase. They also reported that AFGP-8 was retained partially in the ice phase and that it functioned as a weak antifreeze agent. The incorporation of active AFGP into the ice phase has been confirmed by Tomimatsu et al. (*30*), but AFGP-8 has not been found to be incorporated into the ice phase. A greater resistance to exclusion of NaCl in AFGP solutions has been reported also (*46*).

Table VI. Freezing and Melting of Water and Solution of Antifreeze Glycoproteins in Water[a, b]

Temperature Adjustments (°C phase)	*Observed Changes*	
	In Water Containing Ice Crystals	*In Water Solution of 1% Antifreeze Glycoproteins Containing Ice Crystals*
0.0 holding	melt and freeze	crystals melt
−0.1 lowering −0.7 lowering	frozen	crystals do not melt, liquid does not freeze
−0.8 holding	frozen	crystals grow, new crystals form until all solutions frozen
−0.7 raising −0.1 raising	frozen	all frozen, no melting
0.1 holding	melt and freeze	melt

[a] From Feeney and Hofmann (*11*).
[b] Water and antifreeze glycoprotein solution initially was frozen at −3°C and then allowed to melt at +0.1°C until 5–10% of solution remained as ice crystals. The temperature then was adjusted to 0.0°C and periodically lowered and then raised as indicated. The times at each temperature intermediate between freezing and melting were 5-10 min. All observations were made microscopically.

Nature

Characteristics of the Ice Frozen from Solutions of Active Antifreeze. Since the melting temperatures of ice frozen from solutions of active AFGP are normal (i.e., approximately 0°C), the crystal structure of the ice from AFGP solutions would most likely be similar to the crystal structure of pure ice. The similarity of the AFGP ice to normal ice has been reported by Duman and DeVries (*45*), who cited that from x-ray studies the ice from AFGP solutions was ordinary hexagonal ice.

The similarity of the ice from AFGP solutions to ordinary ice also has been confirmed by Raman studies (*30*). All spectra comparisons indicated that the ice from AFGP solutions was ordinary ice. The Raman spectrum in the OH stretch region showed no difference from the spectrum of pure ice. The spectra also showed that the OH spectrum of 1% solutions of AFGP-4 is similar to that of pure water.

Although the ice formed from AFGP-4 solutions appeared to be ordinary ice, disorientation of the crystal and poorly defined ice-solution interfaces have been observed (*30, 47*). When a 1% solution of Glycoprotein 4 was used to grow an oriented *a*-axis single crystal of ice, it was impossible to obtain a reasonable uniform crystal of ice, even with the very slow growth at −4.5°C and the continuous sweeping of the ice-solution interface with fresh glycoprotein solution. The ice that formed showed poor polarization properties with numerous regions of uneven extinction between cross polaroids (*30*).

Characteristics of the Antifreeze Glycoprotein in Solution and in the Ice Phases. Raman spectroscopic studies of AFGP-1 to AFGP-5 frozen into the ice phase or remaining in the liquid phase showed some structural differences that may be useful in a mechanistic interpretation (*30*). Difference spectra is most pronounced in the COH vibrational region.

Direct observation of the AFGP in solution by quasi-elastic light scattering did not show any conformational changes that could be related to function (*30*). Comparison of measurements of diffusion coefficients at different temperatures and at −0.2°C in the presence of ice crystals did not show that any conformational changes in the proteins occurred under conditions where the protein was functioning, i.e., preventing ice-crystal growth at −0.2°C.

Possible Mechanisms of Antifreeze Functions

The Physics of the Freezing Process. A commonly encountered process resulting in low freezing temperatures is supercooling (*7, 8*). Pure water has been known to exhibit freezing temperatures of nearly −40°C because of extensive supercooling. This can be explained by considering the statistical probability that water molecules will cluster and exceed the critical dimension for crystal formation through favorable local enthalpic and entropic conditions. The large degree of supercooling reflects the fact that extensive structural rearrangements occur in the process of freezing from the liquid phase. Seeding pure water with ice-structure promoters usually leads to less supercooling. These substrates stimulate a more favorable ice-lattice bonding environment (*48, 49*). Conversely, colligative depressants function simply by the fact that as they dissolve there is a decrease in the free energy of the total system.

Antifreezing molecules may function as inhibitors of ice nucleation or ice-crystal growth by several means. These noncolligative mechanisms could include the poisoning of nucleation sites which were provided by seeding with heterogeneous nucleators. Another possible mechanism is a change in the properties of water, probably through a restructuring of water by the randomization of the local tetrahedral hydrogen bonding feature of liquid H_2O interaction with the additive. The freezing process would require expenditure of energy in reordering liquid H_2O and subsequently converting it to ice structure. The excess energy needed for ordering could result in the lowering of the freezing point. Finally, the mechanism could be an interaction of the antifreeze agent with H_2O, forming some type of solid solution which has a crystal structure that is not normal ice.

All of the above suggested mechanisms by which these antifreezing biopolymers might function seem highly improbable. Colligative phenomena can be excluded on the basis of the low concentration of the polymer. Colligative effects can contribute only 1/500 of the total lowering of the freezing temperature (*6, 8*). Nucleation-site poisoning probably can be excluded also because AFGP functions in the presence of ice crystals (*11*).

Interactions with liquid water or the ice phase are unlikely on several grounds. Even allowing for a maximum of 10 H_2O molecules per disaccharide group (*29*), a 1% solution of AFGP could bind less than 0.2*M* H_2O, far less than the total H_2O molecular content. Raman study (*30*) of the bulk liquid water with and without AFGP showed that the OH-stretching regions of these two samples are indistinguishable, again suggesting that no new H_2O–AFGP bonds are formed at the functioning temperature.

If AFGP were to enter into some form of solid solution with H_2O in the bulk phase, then the entire structure of such a phase will differ from that of pure ice. Furthermore, the melting temperature will not be that of the pure ice. These features have not been observed (*8, 30, 50*).

A more probable situation can occur if the AFGP molecules form a unique solution with a surface layer of H_2O molecules. Such a layer has been postulated to exist for the pure ice by Fletcher (*51*). For the present case, we postulate that such a solution layer sufficiently alters the growth property of the ice nuclei and yet its presence is unseen by conventional methods. In order to satisfy the observation of normal melting temperature, we further postulate that the phase equilibrium curves of this solution lie within the actual antifreeze functioning temperature range. In this manner, phase separation at temperatures below −0.8°C would lead to continued growth of the previously inhibited ice crystal embryo.

Another possible mechanism of antifreeze function is through interfacial disruption of the growth sites of the subcritical ice nuclei. Such a mechanism has been suggested previously (*25, 47*) and has been elaborated on recently (*7, 8*).

Evidence for an interfacial mechanism must necessarily be based upon the existence of an ice–liquid H_2O interface. We postulate that the existence of a precritical embryonic ice crystal is possible. Consequently the 0°C environment permits the dynamic equilibrium between liquid H_2O molecules ($H_2O(l)$) and nuclei ($H_2O(i)$), where i is the index of nuclei size. In particular:

$$H_2O(l) \rightleftharpoons \sum_{i_{\text{all values}}} H_2O(i)$$

exists. Without AFGP supercooling is a measure of time for development of a nucleus which is supercritical in size $i > i_c$.

The presence of AFGP molecules leads to a further reaction:

$$\mathrm{AFGP} + \sum_{i_{\text{all values}}} \mathrm{H_2O}(i) \rightleftharpoons \sum_{i<i_c} \mathrm{AFGP} \cdot \mathrm{H_2O}(i)$$

Such a reaction will displace the population of subcritical but free nuclei, and thus reduce the possibility of macroscopic ice development.

As the temperature of the total system is lowered, we suggest that a barrier begins to form, making the previously favorable species, AFGP · $H_2O(i)$, less favorable. This barrier may have its origin in the structural mismatch which will be dealt with in the next section. The increasingly unfavorable structural match between AFGP and the specified ice structure leads to the existence of AFGP-free species again. Thermodynamically,

$$\sum_{i<i_c} \mathrm{AFGP} \cdot \mathrm{H_2O}(i) \rightleftharpoons \mathrm{AFGP} + \sum_{i} \mathrm{H_2O}(i)$$

at this stage (-0.8°C), the more favorable reaction:

$$\sum_{i} \mathrm{H_2O}(i) \rightleftharpoons \mathrm{H_2O}(s)$$

can take place. Consequently, normal ice is formed and AFGP has done its share as an equilibrium deterrent against spontaneous ice formation. In studies on the growth pattern of ice in the presence of AFGP at low temperatures under crossed-polaroid examination, the *a*-axis growth is greatly impaired when the active forms of AFGP are present (*30*). This observation suggests that AFGP may be inhibiting ice formation by shielding certain growth sites. AFGP becomes entrapped (*30, 47*) as ice growth continues. Here entrapment means that the ice which is formed is in every way normal ice, but AFGP still is within the ice region. This could occur if the grain boundaries of the ice crystals trap and hold the AFGP molecules within their boundaries. Such an AFGP entrapment picture again is consistent with an interfacial mechanism. However, experimental evidence is not available at this time to describe the orientation of either the hydrophobic or hydrophilic groups at the interface.

The situation is not completely reversible upon warming to the normal ice-melting temperature. Here, the very few AFGP molecules ($10^{-3}M$) cannot interact fully with all of the hydrogen-bonding possibilities which formed the macroscopic ice structure. A few surface-interacting regions do not affect the volumetric effect of normal ice since for a macroscopic crystal, the surface-to-volume ratio is $\sim (N)^{-1/3}$, where N is the number of molecules. As $N \rightarrow \infty$, this ratio tends toward zero. Consequently, such a system will return to 0°C before melting, just like normal ice, and hence the observed hysteresis in the freezing process.

Mechanisms of Action of AFGP and AFP. In Figure 7 the unit-cell directions of the ice I structure have been outlined specifically on top of the tetrahedral bonding structure. It is important to note that the slow growth axis of ice, the *c*-axis, has a unit repeat distance of 7.36 Å when the temperature is close to the pure ice-melting temperature. When the AFGP molecules are in an expanded configuration, their disaccharide groups can be nearly 10 Å apart. Such a distance can readily facilitate coverage of the sites perpendicular to the *c*-axis of the ice crystal. Furthermore, when such a situation occurs, the tripeptides can most readily be forced to expose their methyl hydrophobic groups along the *a*-axis of the ice crystal. Since the *a*-axis is the fast-growth axis for ice-crystal propagation, AFGP in such a disaccharide-bonded condition may be able to prevent both *a*-axis propagation as well as basal-face growth.

Such coverage by AFGP is even more striking when one examines the hexagonal layout of the ice-crystal plane perpendicular to the *c*-axis (Figure 7). One sees that the O–H– bonding distance in H_2O (ca. 2.76 Å) is similar in magnitude to the distance between hydroxyl groups of the disaccharide (ca. 2.86 Å).

With respect to the AFP, DeVries and Lin (*34*) have stated that the 4.5 Å between the hydrophilic amino-acid groups (Asp–Glu) of the α-helical conformation will be able to lay on the ice-basal plane as readily as the sugar-ring structure does. Furthermore, when these groups are hydrogen bonded across more than one plane, the section of Ala–Ala–Ala between hydrophilic groups is again free to be blocking *a*-axis growth in the same manner that the AFGP might function with the Ala–Ala–Thr backbone. Preliminary results from our laboratory indicate that the mechanism of action of AFGP and AFP might be the same. These preliminary results show that mixtures of solutions of AFGP and AFP lower the freezing temperature in the same way as the individual solutions lower it (*37*).

These structural matches which we have displayed are of great interest and importance in our attempt to understand how these systems lower the ice-freezing temperature in a noncolligative manner. In particular, we have emphasized here not only the structural match on the basal planes, but that the methyl groups of the predominant amino-acid residue can indeed play a major role in the antifreezing mechanism.

Vandenheede et al. (*25*) first pointed out possible roles for the hydrophobic methyl groups. They suggested that the AFGP might form clathrate-type inclusion bodies with developing ice crystals and stated

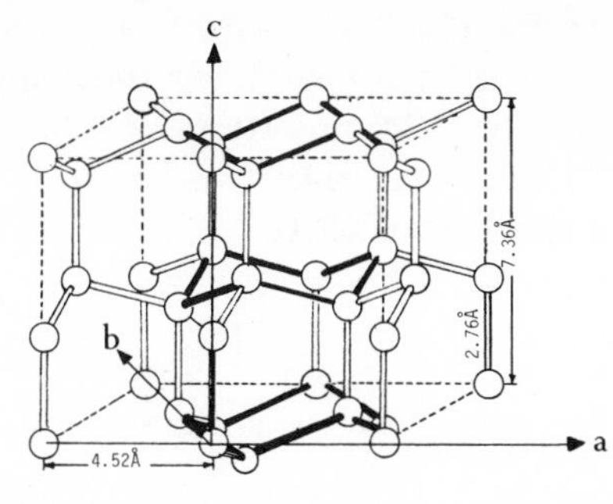

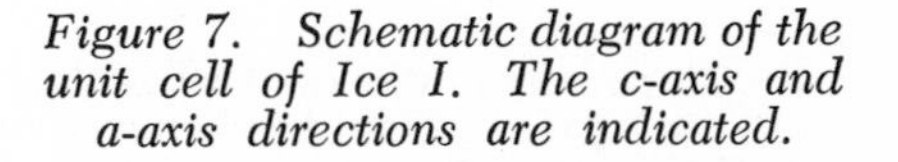
Figure 7. Schematic diagram of the unit cell of Ice I. The c-axis and a-axis directions are indicated.

that the mechanism of action probably involved an interface separating the solid and liquid phases during the growth of ice crystals. Although the complete structure of the several types of AFP are not known at this time, their high alanine content strongly indicates a primary role for the hydrophobic methyl group in their antifreeze action.

In considering how the cooperative potentiation of antifreeze activity can occur between AFGP-1 to AFGP-5 and AFGP-8, a most obvious question is whether the AFGP-1–5 activates the AFGP-8, or whether the AFGP-8 potentiates the function of the AFGP-1 to AFGP-5. A likely model appears to be one in which AFGP-1 to AFGP-5 provides a center around which the smaller AFGP-8 can function. Extending this to a model whereby the AFGP-1 to AFGP-5 functions at an interfacial region between the ice or ice nucleus and the solution, AFGP-8 could somehow fit alongside molecules of AFGP-1 to AFGP-5. Based on this model, AFGP-8 would not be present to any significant degree at the interface in the absence of AFGP-1 to AFGP-5 in supercooled solutions.

Detailed studies of the ice–water interfacial region, in the presence and absence of the mixture, are presently under way. Finally, any consideration of the functional models for antifreeze glycoprotein, which now includes potentiation, also must be tested on AFP. Potentiating substances in the sera of species with AFP probably would be very different from those in the species with AFGP.

Acknowledgment

The authors greatly appreciate the editorial assistance of Clara Robison, Chris Howland, Patty Cotta, and Laura Hayes in the preparation of this manuscript and the National Institutes of Health (Grant No. HD 00122 for R.E.F.) and the National Science Foundation (Grant No. GA-12607 for R.E.F. and BMS 73–06918 for Y.Y.) for their financial support pertaining to this research. Particular appreciation is given to the National Science Foundation for their advice and logistics in research in the Antarctic (*12*) and for their support in research on other proteins from Antarctic species (*52*). The authors are also thankful to the Royal Norwegian Council for Scientific and Industrial Research for their financial assistance and to the Norwegian Institute of Marine Research, Bergen, for their hospitality to R.E.F. and D.T.O. for over three-weeks cruise on their research vessel, the G. O. Sars, in the Fall of 1976.

Literature Cited

1. Kanwisher, J. W. In "Cryobiology; Freezing in Intertidal Animals"; Meryman, H. T., Ed.; Academic: New York, 1966; p 487.
2. Scholander, P. F.; van Dam, L.; Kanwisher, J. W.; Hammel, H. T.; Gordon, M. S. *J. Cell Comp. Physiol.* **1957,** *6,* 747.
3. Leivestad, H. *Spec. Publs. Int. Comm. NW Atlant. Fish.* **1965,** *6,* 747.
4. Ahmed, A. I.; Osuga, D. T.; Feeney, R. E. *J. Biol. Chem.* **1973,** *248,* 8524.
5. DeVries, A. L. In "Biochemical and Biophysical Perspectives in Marine Biology"; Malins, D. C., Sargent, J. R., Eds.; Academic: New York, 1974; Vol. 1, p 289.
6. Feeney, R. E. *Am. Sci.* **1974,** *62,* 712.
7. Yeh, Y.; Feeney, R. E. *Acc. Chem. Res.* **1978,** *11,* 129.
8. Feeney, R. E.; Yeh, Y. *Adv. Protein Chem.* **1978,** *32,* 191.
9. Komatsu, S. K. Ph.D. Thesis, University of California at Davis, 1969.
10. Osuga, D. T.; Feeney, R. E. *J. Biol. Chem.* **1978,** *253,* 5338.
11. Feeney, R. E.; Hofmann, R. *Nature* **1973,** *243,* 357.
12. Feeney, R. E. "Professor on the Ice"; Pacific Portals: Davis, CA, 1974.
13. Scholander, P. F.; Flagg, W.; Walters, V.; Irving, L. *Physiol. Zool.* **1953,** *26,* 67.
14. Gordon, M. S.; Amdur, B. H.; Scholander, P. F. *Biol. Bull.* **1962,** *122,* 52.
15. Scholander, P. F.; Maggert, J. E. *Cryobiology* **1971,** *8,* 371.
16. Hargens, A. R. *Science* **1972,** *176,* 184.
17. Raymond, J. A.; Lin, Y.; DeVries, A. L. *J. Exp. Zool.* **1975,** *193,* 125.
18. Duman, J. G.; DeVries, A. L. *Comp. Biochem. Physiol. B* **1976,** *54,* 193.
19. Duman, J. G.; DeVries, A. L. *Nature* **1974,** *247,* 237.
20. Hew, C. L.; Yip, C. *Biochem. Biophys. Res. Commun.* **1976,** *71,* 845.
21. DeVries, A. L.; Wohlschlag, D. E. *Science* **1969,** *163,* 1073.
22. DeVries, A. L.; Komatsu, S. K.; Feeney, R. E. *J. Biol. Chem.* **1970,** *245,* 2901.
23. Komatsu, S. K.; DeVries, A. L.; Feeney, R. E. *J. Biol. Chem.* **1970,** *245,* 2909.
24. DeVries, A. L.; Vandenheede, J.; Feeney, R. E. *J. Biol. Chem.* **1971,** *246,* 305.
25. Vandenheede, J.; Ahmed, A. I.; Feeney, R. E. *J. Biol. Chem.* **1972,** *247,* 7885.
26. Osuga, D. T.; Ward, F. C.; Yeh, Y.; Feeney, R. E. *J. Biol. Chem.* **1978,** *253,* 6669.
27. Vandenheede, J. R. Ph.D. Thesis, University of California at Davis, 1972.
28. Morris, H. R.; Thompson, M. R.; Osuga, D. T.; Ahmed, A. I.; Chan, S. M.; Vandenheede, J. R.; Feeney, R. E. *J. Biol. Chem.* **1978,** *253,* 5155.
29. Osuga, D. T.; Feeney, R. E., unpublished data, 1977.
30. Tomimatsu, Y.; Scherer, J. R.; Yeh, Y.; Feeney, R. E. *J. Biol. Chem.* **1976,** *251,* 2290.
31. Franks, F.; Morris, E. R. *Biochim. Biophys. Acta* **1978,** *540,* 346.
32. Mulvihill, D. M.; Geoghegan, K. F.; Yeh, Y.; DeRemer, K.; Osuga, D. T.; Ward, F. C.; Feeney, R. E. *J. Biol. Chem.* **1979,** submitted for publication.
33. Hew, C. L.; Yip, C.; Fletcher, G., personal communication, 1977.

34. DeVries, A. L.; Lin, Y. *Biochim. Biophys. Acta* **1977**, *495*, 388.
35. Ananthanarayanan, V. S.; Hew, C. L. *Biochem. Biophys. Res. Commun.* **1977**, *74*, 685.
36. Raymond, J. A.; Radding, W.; DeVries, A. L. *Biopolymers* **1977**, *16*, 2575.
37. Osuga, D. T.; Ward, F. C.; Yeh, Y.; Hew, C.; Feeney, R. E. "Abstracts of Papers"; 1978 Pacific Slope Biochemical Conference, Santa Barbara, CA, Session IV, p 34.
38. Duman, J. G.; DeVries, A. L. *J. Exp. Zool.* **1974**, *190*, 89.
39. Fletcher, G. L.; Campbell, C. M.; Hew, C. L. *Can. J. Zool.* **1978**, *56*, 109.
40. Ananthanarayanan, V. S.; Hew, C. L. *Nature* **1977**, *268*, 560.
41. Feeney, R. E.; Osuga, D. T.; Ward, F. C.; Rearick, J. I.; Glasgow, L. R.; Sadler, J. E.; Hill, R. L. "Abstracts of Papers," 176th National Meeting, ACS, Sept. 1978; BIOL 13.
42. Ahmed, A. I.; Feeney, R. E.; Osuga, D. T.; Yeh, Y. *J. Biol. Chem.* **1975**, *250*, 3344.
43. Ahmed, A. I.; Yeh, Y.; Osuga, D. T.; Feeney, R. E. *J. Biol. Chem.* **1976**, *251*, 3033.
44. Shier, W. T.; DeVries, A. L. *Biochim. Biophys. Acta* **1972**, *263*, 406.
45. Duman, J. G.; DeVries, A. L. *Cryobiology* **1972**, *9*, 469.
46. Lin, Y.; Raymond, J. A.; Duman, J. G.; DeVries, A. L.; *Cryobiology* **1976**, *13*, 334.
47. Raymond, J. A.; Lin, Y.; DeVries, A. L. *Proc. Nat. Acad. Sci. USA* **1977**, *74*, 2589.
48. Fletcher, N. H. "The Chemical Physics of Ice"; Cambridge Univ.: Cambridge, England, 1970; p 26.
49. Zettlemoyer, A. C. *J. Colloid Interface Sci.* **1968**, *28*, 343.
50. Raymond, J. A. Ph.D. Thesis, University of California at San Diego, 1976.
51. Fletcher, N. H. *Philos. Mag.* **1968**, *18*, 1287.
52. Feeney, R. E.; Osuga, D. T. *Comp. Biochem. Physiol. A* **1976**, *54*, 281.

RECEIVED June 16, 1978.

6

Chemical Reactions in Proteins Irradiated at Subfreezing Temperatures

IRWIN A. TAUB and JOHN W. HALLIDAY

Food Engineering Laboratory, U. S. Army Natick Research and Development Command, Natick, MA 01760

MICHAEL D. SEVILLA

Department of Chemistry, Oakland University, Rochester, MI 48063

The course of chemical reactions in irradiated proteins is determined by factors that influence the reactivity of the primary free radicals, the kind of protein radicals formed, and the decay of these protein radicals to stable products. To understand these reactions, basic radiation chemical concepts are considered, chemical changes in several representative proteins irradiated under different conditions are compared, and results from optical and electron spin resonance studies on model systems are presented. Among the reactions described are those involving cation, anion, and α-carbon radicals of amino acids and peptides. Analogous reactions common to proteins are then summarized. These mechanistic considerations have important implications for the irradiation of hydrated muscle proteins at −40°C and for radiation sterilization of foods.

The chemical reactions occurring in a protein system exposed to ionzing radiation are affected by several factors. The nature of the protein, its state of hydration, the phase and temperature of the system, and the presence of reactive compounds are particularly important factors. Proteins containing disulfide groups, metal ions, or large proportions of aromatic or heterocyclic amino acids will undergo reactions differing from those without these constituents. Proteins that are hydrated

0-8412-0484-5/79/33-180-109$8.00/0

undergo different chemical and physical processes than proteins that are desiccated. Most important, however, is the physical state: fluid solutions are much more susceptible to radiation induced changes than frozen aqueous systems. Temperature, particularly as it affects the viscosity of the medium, can change the chemical consequences markedly. Solutes in these systems, such as oxygen or metal ions, can alter the course of the reactions as well.

Irradiation initiates a series of reactions as a consequence of electrons being ejected and bonds in the protein being ruptured. Various ionic and free radical intermediates are generated that ultimately become stabilized by the formation of covalently bonded compounds. Reactions of the primary radicals can generate other radicals and these in turn combine to form the final products. Any factor that affects the rates and routes of these intermediates will influence which products are formed. For the proteins of interest here, the final effects might be observable as valence change, deamination, decarboxylation, disulfide loss or formation, chain degradation or aggregation, or modification of individual amino acid moieties.

An understanding of these reactions and how they are affected, particularly by irradiation at subfreezing temperatures, can be achieved by considering certain basic concepts and major experimental observations. Consequently, the basic radiation chemical concepts, the techniques used to discern intermediate species and their eventual products, the major findings on such proteins as myoglobin and myosin, and the generalized reactions of primary and secondary radicals as gleaned from studies on amino acid and peptides will be considered herein. The implications of these findings, though generally relevant to radiation biology, redox processes in biochemistry, and protein dynamics, will be discussed in relation to low temperature radiation sterilization of high protein foods.

Basic Radiation Chemical Concepts

Energy Deposition and Free Radical Distribution. The interaction of penetrating gamma rays or high energy electrons with the valence shells of the atoms comprising the molecules of the condensed medium results in energy being deposited in the medium. These electrons can be produced directly in machine sources such as a linear accelerator or a Van de Graaff accelerator, and would have energies typically in the range of 2–10 MeV. Gamma rays, such as those from cobalt-60 or cesium-137 sources, as a consequence of interacting via the Compton process produce energetic electrons as well. Because of its charge, it is the electron that is particularly effective in excitation and ionization processes.

By successive ionization acts, the primary electron gradually becomes degraded in energy and concomittantly produces secondary electrons, also capable of ionizing the molecules. Ultimately, these primary, secondary, and tertiary, etc. electrons can no longer cause ionizations, and lose their remaining energy via electronic, vibrational, and rotational excitation. An energetically degraded electron might be drawn back to the positive ion formed upon ionization, might be trapped if the medium is polar and become solvated, or might react with an impurity of high electron affinity. Some of the excited molecules formed directly or produced upon electron-positive ion reaction will dissociate into free radicals. The overall effect is the initial formation of ions and free radicals nonuniformly distributed in regions called spurs along the track of the ionizing particle.

The distribution of these primary species and their eventual fate is determined by the nature and state of the medium. If the viscosity is extremely high as in solids or glasses, the distribution remains nonuniform and the reactions that occur involve species within the same, or closely related, spur. If the viscosity is low, such as in a fluid system, these species tend to diffuse apart and lead to a uniform distribution throughout the medium. In this case, the reactions conform to kinetic laws for a homogeneous system.

The yield of any species formed as a direct consequence of the energy being absorbed by the component molecules is given in terms of a G-value, which is defined as the number of ions, free radicals, or excited molecules formed for every 100 eV of energy absorbed. G-values may also be given for the stable products that are eventually formed.

Direct Effect on Water. For water, radiolysis leads to the formation of species with rather special features that have been the subject of considerable investigation (*1, 2, 3*). Equation 1 describes the overall effect:

$$H_2O \xrightarrow{\gamma, e^-} (H_2O^+), e_{aq}^-, OH\cdot, H_3O^+, H\cdot, H_2, H_2O_2 \qquad (1)$$

The molecular ion of water is shown in parenthesis because it rapidly converts to $OH\cdot$ and H_3O^+. The solvated electron (*2*), corresponding to an electron bound to several water molecules in a fluid system, is designated here as e_{aq}^-, but will also be denoted as e_s^- for an electron bound in other polar media. It is highly mobile, has a broad, intense absorption spectrum with a maximum at 720 nm, and is a powerful reductant. The hydroxyl radical, $OH\cdot$, is also very mobile and is a strong oxidant; it exhibits a weak absorption in the 240 nm region. The hydrogen atom, $H\cdot$, is a reductant and exhibits only a weak absorption in the ultraviolet region. All three radicals have been detected and characterized by virtue of their electron spin resonance (ESR) spectra. Evidence for an unsolvated electron is extensive and is based on chemical and conductometric measurements, which will be mentioned below. G-values for the e_{aq}^-, $OH\cdot$, and $H\cdot$ uniformly distributed in water at room temperature are 2.8, 2.7, and 0.55, respectively; G-values for these same species in ice at −5°C are approximately 0.3, 1.0, and 0.7, respectively.

For proteins or other organic constituents, radiolysis presumably leads to analogous species, although these have not been unequivocably established. Equation 2 describes the overall effect:

$$PH \xrightarrow{\gamma, e^-} (PH^+), e^-, P\cdot, PH_2^+, H\cdot, H_2, P_2 \qquad (2)$$

The generalized protein molecule is shown as PH. The e^- is left unspecified; its fate would depend on the medium and other conditions. As will be described later, there are several free radicals that have been detected, but the general designation $P\cdot$ is all that is needed here.

Effect of Phase, Temperature, Solutes, and Dose on the Radiolysis. Since the primary radical species must diffuse to other radical species or molecules in the system to transfer or share electrons and become stable, the phase and temperature, because they affect viscosity, determine which reactions will occur. Those reactions involving the primary species and the molecules not directly affected by the radiation give rise to new species—secondary radicals—which are considered as being an indirect consequence of the radiolysis. In fluid aqueous systems, e_{aq}^-, $OH\cdot$, and $H\cdot$ readily react with each other and with solutes, even those present at low concentrations ($\sim 10^{-4}$ *M*). In frozen systems, such reactions are impeded by the rigidity of the medium. Indirect consequences are limited to systems with very high solute concentrations (~ 1 *M*) and at temperatures near 0°C, both factors contributing to formation of amorphous or fluid-like regions. Studies of the reaction of e_s^- with various solutes in polycrystalline systems at −40°C have shown that G-values for products are only about 1% to 10% of the corresponding values found for fluid systems (*4, 5, 6*). When chloroacetic acid was used as a probe for the electrons, the G-value for Cl^- formation increased from approximately 0.03 to 0.8 as the chloroacetic acid concentration was increased from 10^{-2} *M* to 1 *M*. Similarly, studies of reactions of $OH\cdot$ in polycrystalline ice at −40°C have shown that this species is even more restricted than e_s^- in its ability to migrate and react (*4*). These experiments involved using ferrocyanide as a probe for $OH\cdot$. It has been estimated that only 4% of the available $OH\cdot$ can be scavenged using

0.5 *M* ferrocyanide, the highest practical concentration obtainable. Consequently, the overall chemical change in a frozen or solid system is significantly reduced by limiting the indirect formation of other species.

The course of primary species reactions in a fluid system, which is dependent upon the rate constants for reaction (see Table I) and the concentrations of reactive solutes, determines the nature of the indirect effects. The solutes could be introduced directly or could be formed as a consequence of the radiolysis. For systems in which homogeneous kinetic laws apply, the predominant reaction is determined by competition principles: the reaction involving the highest product of rate constant *k* times concentration predominates. If a reactive product is formed in the radiolysis with a reasonably high G-value, it will tend to compete as the dose is increased.

Table I. Rate Constants for Reaction of Primary Water Radicals with Some Amino Acids and Peptides in Fluid Aqueous Solutions[a]

	k, M^{-1} s^{-1} *(pH)*		
Amino Acid/Peptide	e_{aq}^{-}	*OH·*	*H·*
Glycine	8.2×10^{6} (6.4)	1.6×10^{7} (5.2)	9×10^{4} (7)
Glycylglycine	3.7×10^{8} (6.4)	4.4×10^{8} (5.2)	2.6×10^{6}
Alanine	5.9×10^{6} (6.4)	4.7×10^{7} (6)	2.9×10^{5}
Lysine	2×10^{7} (7)	6.0×10^{8} (2)	1.6×10^{6}
Arginine	1.8×10^{8} (6)	3.5×10^{9} (6.5–7.5)	4.9×10^{6}
Aspartic acid	1×10^{7} (7.3)	2.1×10^{7} (6.8)	2.9×10^{6} (7)
Histidine	6×10^{7} (7)	5.0×10^{9} (6–7)	2.5×10^{8} (7)
Phenylalanine	1.5×10^{8} (6.8)	6.6×10^{9}	8.0×10^{8}
Tryptophan	4.0×10^{8} (6.8)	1.4×10^{10} (6.1)	2.3×10^{9}
Methionine	3.5×10^{7} (6.0)	—	—
Cysteine	8.7×10^{9} (6.3)	—	4×10^{9}
Cystine	1.3×10^{10} (6.1)	—	8×10^{9}

[a] From Refs. *2, 3, 52, 56.*

Major Physicochemical and Analytical Techniques

Techniques for Examining and Characterizing Irradiated Proteins. Several techniques have been employed to detect the presence of intermediate species in the irradiated protein or to characterize the changes brought about in its structure or composition. Optical techniques have been used both for detecting intermediates as well as for determining final changes. Because only species with unpaired electrons can be detected by ESR, this technique has been applied for detecting intermediates directly or, most recently, for discerning their nature after trapping with stable free radicals (7). Among the more standard biochemical approaches for characterizing the permanent modifications in the protein are the electrophoretic methods, chemical analyses of products or constituent moieties, and several structure-related methods. Electrophoresis, isoelectric focusing, and chromatographic separations are useful for discerning changes due to loss or modification of moieties that influence size, shape, and charge on the protein. Amino acid analyses, SH group analysis, and determination of amide and fatty acid products pertain to similar modifications. Experiments involving enzymatic activity, sedimentation rates, and binding of radioactive labels also pertain to structural alterations or specific moiety modifications. In general, the analysis of permanent changes is relatively insensitive, and very high doses, often in the range of 3000 kGy (1 Mrad = 10 kGy), must be used. Considerably greater sensitivity is available and a more direct understanding of the chemistry is possible if one uses the techniques recently developed for studying the intermediates.

Detection of Transient (Short-Lived) Intermediates. By pulse irradiating (*8*) a system, it is possible to detect the presence of short-lived intermediates and to characterize their structural and kinetic properties. The technique involves the use of electron accelerators capable of delivering a high dose to a system in a time that is short compared to the lifetime of the species being studied. Typically, such machines deliver doses of about 200 Gy (1 krad = 10 Gy) in a square wave pulse lasting about 1 μs. For systems in which intermediate is formed with a G-value of 3, the instantaneous concentration after the pulse is 6×10^{-5} *M*. If the species has a relatively high extinction coefficient ($\sim 10^3\ M^{-1}\ cm^{-1}$), it can be monitored with fast spectrophotometric techniques. This optical approach has been used primarily for aqueous solutions; but it has also been applied to aqueous glasses (*9*). Where the magnetic resonance properties of the radicals do not lead to broad lines, ESR can also be used (*10, 11, 12*). Provided a particular species has a reasonably high or sufficiently different conductivity, electrochemical detection is especially effective for charged intermediates (*13*). A recent variation on these techniques has been described by Warman (*14*), in which microwave devices are used to detect high-mobility unsolvated electrons in ice. Although not considered a pulsed or fast reaction approach, the technique developed by Eiben and Fessenden (*15*), involving continuous irradiation and in situ ESR detection, is especially effective. With this technique, radicals that have attained a requisite steady-state concentration can be examined, and a substantial catalogue of spectra for free radicals in fluid aqueous solutions has been compiled.

Detection of Trapped or Stabilized Intermediates. In systems of high viscosity, the free radicals of interest in protein radiolysis can be immobilized and examined with standard optical or ESR techniques. Transparent media must be used for examining the radicals optically and for generating them photolytically (*16*). Aqueous glasses made from hydroxide, perchlorate, or ethanediol solutions are used at low temperatures ($< -90°C$). For a glass equimolar in water and ethanediol and containing a solute of interest, the procedure involves irradiating at less than $-130°C$ with γ rays, photobleaching using visible light the electrons that are trapped in the matrix, and then scanning the spectrum of the species derived from the electron-solute reaction. For generating the radicals photolytically and examining them with ESR, the procedure involves either direct photoionization (*17*) of the solute or first photoejecting electrons from ferrocyanide ion with uv light and then photobleaching the trapped electrons to obtain the desired electron-solute reaction (*18, 19, 20*). Opaque media such as polycrystalline ice plugs can be used if the radicals are generated radiolytically and then examined with ESR as described. In systems such as powders or crystals, the radicals formed upon γ irradiation are also relatively immobile and can be studied (*21*). Since lowering the viscosity in all of these cases allows various rotational processes and diffusional processes to take place, different radicals can be formed and observed as the viscosity and/or temperature is changed. These techniques have been used to identify and characterize many of the free radicals from proteins, amino acids, and peptides that will be discussed here.

Major Effects on Representative Proteins

Before summarizing the reactions generally occurring in proteins irradiated at low temperatures, it is instructive to review some of the major observations that have been made for certain representative proteins. Those selected for illustration are myoglobin, ribonuclease (RNase), the myofibrillar proteins myosin and actomysin, and gelatin. Wherever possible, comparisons will be made between results for fluid and frozen systems.

Myoglobin: A Representative Metallo-protein. Because the iron in myoglobin can exist in two stable valence states, +3 for metmyoglobin (metMb) and +2 for deoxymyoglobin (Mb), the radiation chemistry of myoglobin in fluid aqueous solution is dominated by oxidation and reduction reactions. Accordingly, Satterlee (*22*), who was prompted by the work of others (*23, 24*) to study color changes in irradiated meats and myoglobin solutions, showed by proper selection of conditions that $OH\cdot$ was not responsible for the reduction of metMb. Subsequently, Simic and coworkers (*25, 26*) and Shieh and coworkers (*27, 28*) systematically examined the pertinent reactions, and demonstrated that metMb can be reduced by a variety of free radicals. Reaction 3 applies for both native and denatured metMb:

$$e_{aq}^{-} [CO_2^{-}\cdot, (CH_3)_2\dot{C}OH] + \text{metMb} \rightarrow \text{Mb} \quad (3)$$

Rate constants for the reduction by e_{aq}^{-} and $CO_2^{-}\cdot$ (a reductant derived from formate ion) are 2.5×10^{10} and 2.0×10^{9} $M^{-1}s^{-1}$ respectively, which approach diffusion-controlled limits. The reduction of metMb by $(CH_3)_2\dot{C}OH$ indicated in Figure 1 is typical, being linearly dependent upon dose until almost all the metMb is depleted. In this experiment e_{aq}^{-} is converted to $OH\cdot$ by reaction 4 which, in turn, is converted to $(CH_3)_2\dot{C}OH$ by reaction 5:

$$e_{aq}^{-} + N_2O \xrightarrow{H^+} N_2 + OH\cdot \quad (4)$$

$$OH\cdot + (CH_3)_2CHOH \rightarrow (CH_3)_2\dot{C}OH + H_2O \quad (5)$$

Conversely, the divalent ion in oxymyoglobin (MbO_2) is readily oxidized to metMb, as reaction 6 indicates:

$$OH\cdot + MbO_2 \rightarrow \text{metMb} + OH^{-} \quad (6)$$

Since metMb is not easily (if at all) further oxidized to a higher valent iron, its reaction with $OH\cdot$ does not appear to affect the iron, but apparently involves the amino acid moieties on the surface of the protein. Their involvement is supported by the formation of dimeric metMb at high doses.

If the irradiation is carried out in a frozen system, however, these reactions are either minimized or eliminated. Reduction by e_s^{-} is possible but restricted. Experiments involving metMb in water–ethanediol glasses at −130°C show that if electrons could interact with this compound, reduction would take place. Figure 2 shows the development of Mb in the glass following irradiation and photobleaching of trapped electrons, which moves the electron to the immobile protein. The implication of these results is confirmed by experiments (*29*) on polycrystalline solutions of metMb irradiated over a range of temperatures from −80 to 0°C. The G-value for reduction, which is 3.1 at room temperature, is only 0.07 at −45°C. There is a slight increase in this value in going from −45 to 0°C, which is consistent with other studies on frozen ices (*4, 5*). Reaction with $OH\cdot$ is even less probable, and chromatographic analyses of metMb irradiated at −45°C to doses as high as 80 kGy show no high molecular weight myoglobin oligomers.

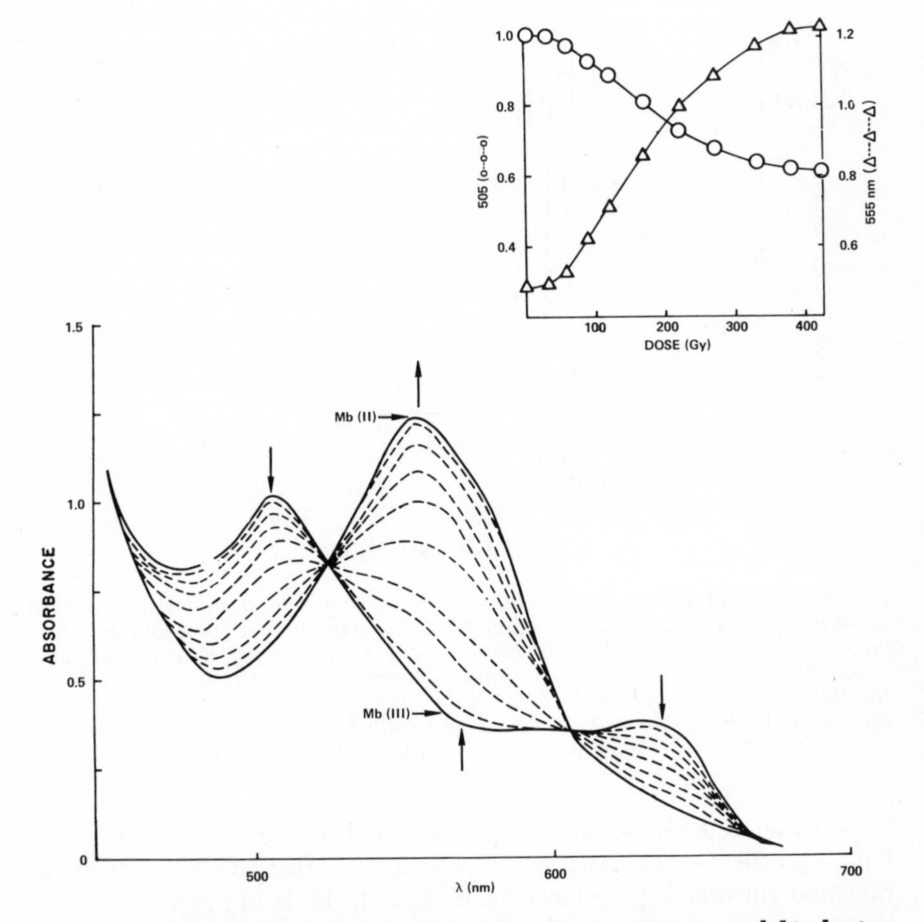

Figure 1. Spectral changes associated with reduction of metmyoglobin by isopropanol radicals at 20°C. Solution contained 1×10^{-4}M *metmyoglobin, 0.05*M *isopropanol, and* 10^{-2}M *phosphate buffer (pH = 7.5), and was saturated with* N_2O. *Spectrum before irradiation is designated by Mb(III), and after completion of irradiation, by Mb(II); spectra for intermediate cases are dotted.* (Inset) *Plot of absorbance at 555 and 630 nm vs. dose.*

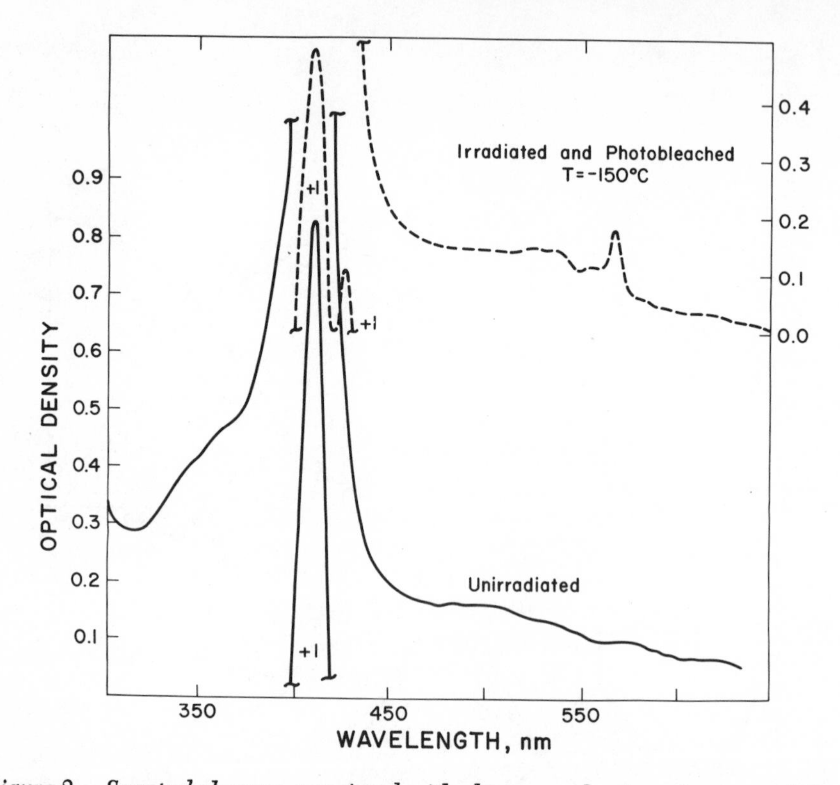

Figure 2. Spectral changes associated with electron reduction of metmyoglobin in an ethanediol–water glass at −150°C. Glass was formed from an equimolar mixture of ethanediol and water containing 1 g/liter of metMb, irradiated to approximately 8 kGy at −196°C, and exposed to visible light to bleach the trapped electrons at −150°C. Solid curve corresponds to the unirradiated system; dotted curve, which is displaced vertically for clarity, corresponds to the final spectrum. Change of scale is indicated by the +1 designation.

RNase: A Representative Enzyme. Both H· and OH· radicals in fluid, aqueous solutions react with RNase in ways that lead to inactivation. Mee and coworkers (*30*) have found that H· leads to aggregation, and that cystine, methionine, and tyrosine are mainly affected. Adams and coworkers (*31*), investigating the specificity of free radical reaction with RNase, found that OH· and $Br_2^{-}\cdot$ are effective inactivators because they rapidly react with histidine, which is associated with the active site.

Structural changes can be discerned in RNase irradiated in solution or in the dry state. Delincée and Radola (*32*) could show from isoelectric focusing measurements on irradiated 0.1 and 1% RNase solutions that several active components having lower isoelectric points are produced. Their gel chromatographic results showed that aggregates are formed stepwise from monomer to dimer to higher polymers. Experiments on dry

RNase gave similar results but required much higher doses: the gel chromatographic pattern obtained from a 0.1% RNase solution irradiated to 0.5 kGy could be matched by irradiating a dry preparation to 150 kGy. Friedberg (*33*) also studied dry RNase and found that its apparent average molecular weight is increased upon irradiation to 300 kGy. He concluded that the structure of globular proteins imposes constraints favoring combination of radicals.

The effectiveness of irradiation in inactivating dry RNase depends on the temperature, as shown by Fluke (*34*). He determined the doses needed to reduce the activity to 37% of its original value (D_{37}) at temperatures from −160 to 182°C. Arrhenius-type plots of D_{37}^{-1} vs. T^{-1} are found to be nonlinear and were interpreted in terms of a temperature-independent term and two temperature-dependent terms. The inactivation becomes relatively effective above about 60°C.

The significance of structure and temperature on the types of free radicals formed in irradiated, dry RNase has been demonstrated by Riesz and White (*35*). Using a labelling technique in which tritium becomes distributed among the amino acids as a result of free radicals reacting with tritiated hydrogen sulfide, HST, they could show different sets of radicals and reactions occurring at −78°C and at 25°C. They found, for example, that the distribution of tritium in native RNase at −78°C is the same as in denatured RNase at 25°C, but is different than in native RNase at 25°C. Apparently, at the lower temperature the radicals are formed randomly and independent of conformation, but their conversion at higher temperature is specific and depends on the particular protein conformation.

Actomyosin and Myosin: Representative Myofibrillar Proteins. The radiolytic effects on solutions of either actomyosin or myosin above and below the temperature of freezing cannot easily be compared. So few studies on these proteins in fluid systems have been conducted, presumably because of their low solubility. However, results from a study by Coelho (*36*) on actomyosin in solutions containing $CaCl_2$ and $MgCl_2$, in which both activity (ATPase) and chemical analyses were made, can serve as reference. Of interest here was whether any effect of radiolysis on enzyme activity could be correlated with chemical and/or structural changes. Coelho found that as the dose was increased, the activity increased up to a maximum (achieved at 2.5 kGy) and then decreased monotonically thereafter. The number of accessible and/or buried SH groups, in contrast, decreased steadily with increasing dose. Ultracentrifugation measurements were also made and these showed that the developing reticulation might be related to the changes in enzyme activity. There is no definite conclusion that can be reached about the reactions responsible for either the change in activity or the overall indications of chemical change. Since Cl^- was used in high concentrations, $Cl_2^-\cdot$, formed through reaction 7, would have been the predominant reactive species:

$$OH\cdot + Cl^- \overset{Cl^-}{\rightleftarrows} Cl_2^-\cdot + OH^- \qquad (7)$$

Consequently, many amino acid sites on the actomyosin would have been susceptible to oxidation by this radical.

Considerably more information on the radiolysis of frozen, hydrated actomyosin and myosin has recently become available (*5, 6, 37, 38*) showing that these proteins are relatively stable towards radiolytic decomposition. The experimental procedures involved isolating the proteins from prerigor beef, dialyzing to remove Cl^-, denaturing some of the samples with a mild heat treatment (70°C, 30 min), preparing the samples either as suspensions or precipates, freezing them, irradiating them, analyzing some by ESR or electrophoresis, and determining the amino acid composition in others after desiccating and hydrolyzing. In this way the influence, if any, of structure could be ascertained and the formation and fate of free radicals as they relate to protein degradation or aggregation and to amino acid modification could be discerned.

ELECTRON SPIN RESONANCE RESULTS. Irradiation of both actomyosin and myosin in either the native or heat-denatured state at −40°C leads to the formation of the same spin centers. Figure 3 shows a high resolution spectrum for actomyosin. It is characterized by a broad doublet signal, having a peak-to-peak separation of 28 gauss, and eight other, weaker lines extending about 130 gauss. The absence of signals attributable to N-centered or S-centered radicals rules out contributions from other than C-centered radicals. Based on comparisons with spectra from dipeptide radicals (to be discussed below), this spectrum is attributed to contributions primarily from different α-carbon radicals on the peptide backbone and secondarily from side chain radicals. Signals for backbone radicals of the type

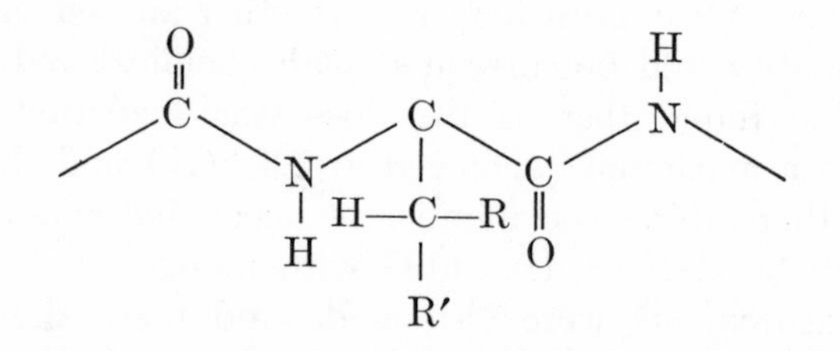

correspond to interaction of the unpaired electron with protons on the associated methylene carbon. Depending on the nature of the R and R′

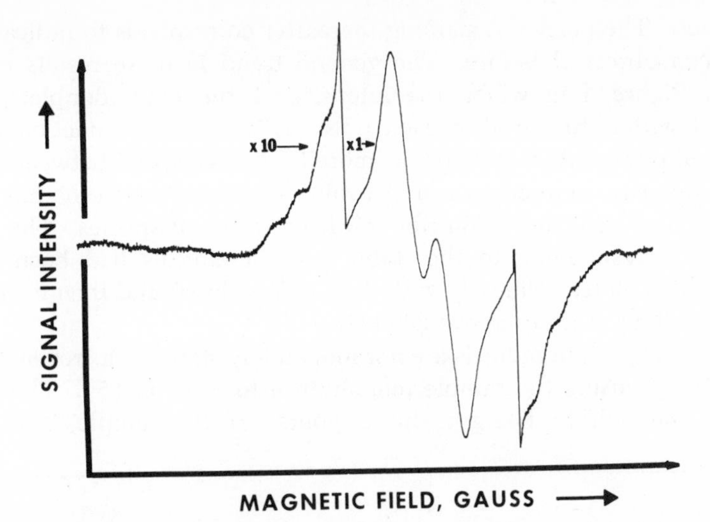

Figure 3. High resolution electron spin resonance spectrum of an irradiated, frozen suspension of actomyosin. Sample was irradiated to 50 kGy at −40°C and spectrum was recorded at −40°C. The central portion of the spectrum corresponding to a gain of ×1 shows the doublet feature with a shoulder on either side. At a gain of ×10, an additional three lines are clearly discernible on both the high and low field portions of the spectrum.

groups, sets of doublet and quartet lines would be produced. Of the side group radicals that could contribute, analyses show that the H atom adduct of the phenylalanine ring

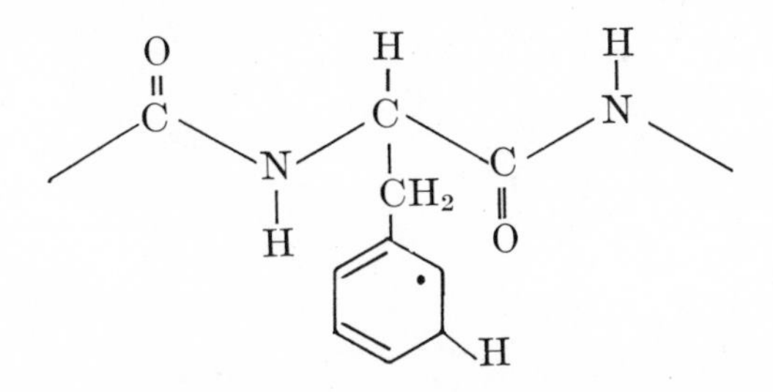

is responsible for the extreme lines and accounts for about 4% of the total radicals. In general, the spectra obtained are very similar to those obtained from other proteins (*39, 40*).

On the basis of conventional ESR measurements, the radicals observed at −40°C appear stable at this temperature. Such backbone radicals on so large a protein with two helically entwined main chains would be immobile in this matrix. More recently, however, pulse ESR experiments on *beef muscle* samples (*41*) indicate that a portion of the radicals formed at −40°C do react within several minutes following irradiation. The portion remaining thereafter corresponds to radicals that have been observed before. The general trend in these results can be seen in Figure 4 in which the intensity of the main doublet signal, recorded within 10 sec after each pulse, is shown as a function of the number of pulses. At certain points more time has elapsed between pulses and the decrease in intensity is noticeable. The curvature in the intensity–dose plot also indicates a contribution from transient species. The signal intensity corresponding to the stable protein radicals had been found previously to increase linearly with dose at low doses, and then to reach a saturation limit at doses above 200 kGy.

That the protein radicals are not indefinitely stable in a frozen system is shown by raising the sample temperature to −10°C (5). The decay is slow, the halflife being about 8 hours. If the sample is thawed,

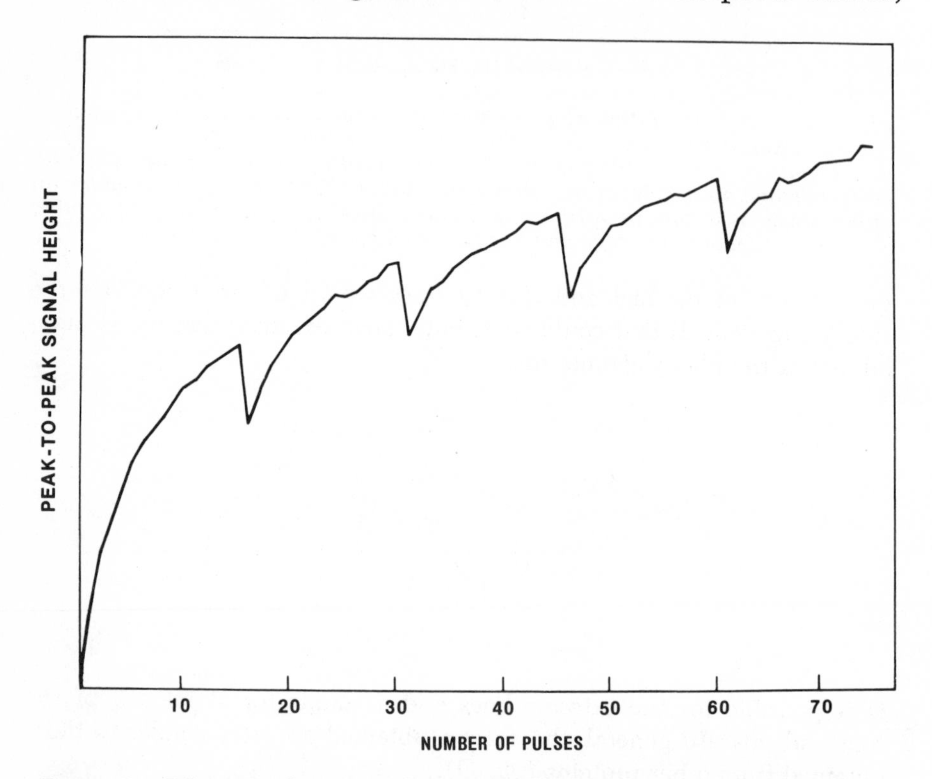

Figure 4. Effect of pulses of irradiation on the ESR signal intensity for radicals in beef at −40°C. Beef samples contained NaCl and tripolyphosphate and had been heat treated to approximately 73°C. Signal intensity corresponds to the peak-to-peak height of the two most intense lines in the spectrum (38). Each spectrum was scanned within 10 seconds after each 5 μsec pulse from a 10 MeV linear accelerator. After every 15 pulses, several minutes elapsed before the pulsing was reinitiated. The dose per pulse was approximately 1 kGy.

refrozen, and re-examined within 30 min, no signal can be detected. Reaction, therefore, is possible provided that certain relaxation processes or short translational motions are promoted at the higher temperatures in the ice. Reaction readily occurs when the constraints of the medium are removed, that is, upon thawing.

SOLUBILITY AND ELECTROPHORETIC RESULTS. Irradiation of these proteins at −40°C over a dose range of about 200 kGy causes a decrease in their overall solubility that is associated with noncovalent aggregation of the main myosin chains (*42*). This decrease in the amount of protein extracted into a solution 8 *M* in urea and 10^{-3} *M* in dithiothreitol or β-mercaptoethanol is nonlinear with dose and reaches a limit at about 80 kGy. The loss in soluble protein is matched by the recoverable protein in the residue. Electrophoretic separations (using SDS and 5% polyacrylamide gels) of extracts of samples receiving successively higher doses show no new bands, no increase in the low molecular weight fragments, a slight development of a very diffuse region centered at about 100,000 daltons, and the gradual loss of the 210,000–250,000 dalton main chain bands. Solubilizing the residue with a guanidine hydrochloride–urea mixture and dialyzing the solution against urea make it possible to examine the protein pattern. Electrophoresis of this material (also with SDS) shows that the main myosin chains have been retained intact. Despite the high doses, no significant aggregation or degradation of actomyosin or myosin occurs under these conditions.

AMINO ACID ANALYSIS RESULTS. Irradiations for these analyses were done with considerably higher doses, 0–400 kGy, to compensate for the limitations in detecting small changes in amino acid composition. All samples were desiccated and then hydrolyzed with *p*-toluenesulfonic acid. The number of residues determined for a particular amino acid was normalized per 1000 residues of all amino acids analyzed. Plots of this ratio for each amino acid against dose are straight lines with zero slope. Such results indicate that, within the limits of reproducibility of about 2%, none of the amino acids is discernibly modified (*37, 42*) under these conditions.

Gelatin: Additional Comparisons. Several experiments on gelatin irradiated as a gel and in a dry system provide an interesting comparison of medium effects.

Stein and co-workers (*43*) irradiated solutions of gelatin ranging in concentration from 1% to 20% and containing 4×10^{-4} *M* of ferricyanide ion at 25°C. Reaction of the gelatin free radicals with ferricyanide would produce ferrocyanide. G-values for this reduction in nitrogen-purged, 1% and 10% gelatin solutions are approximately 2.6 and 5.3, respectively. Since the gels become "stiffer" at higher gelatin concentrations, reaction between gelatin radicals becomes less likely and reduction of the more freely diffusing ferricyanide ion predominates.

Bachman and coworkers (*44*), using ESR techniques, monitored the lifetime of free radicals in dry gelatin irradiated to approximately 50 kGy and stored in air at 20°C. Unlike the radicals in the gel, these are relatively immobile, and required several days to decay appreciably. Some were still detectable after about a month. The work and Friedberg and coworkers (*45*) indicates that the formation of these gelatin radicals in the dry state involves rupture of peptide bonds and that their decay does not involve combination of long chains. They found a decrease in the viscosity of solutions made from dry gelatin irradiated to a dose of 155 kGy.

Other Proteins. In general, the findings described above for these proteins are common to many proteins. More specific information about other proteins can be obtained from available reviews (*46, 47, 48, 49*).

Formation and Conversion Reactions of Amino Acid and Peptide Radicals

Some understanding of the chemical reactions responsible for changes observed in proteins irradiated in fluid and frozen systems can be achieved by considering results from studies on amino acids and peptides. For the sake of simplicity, these reactions are classified according to whether formation of the radical formally involves donating or accepting an electron, whether formation or decay involves conversion from or to another radical, or whether decay involves bimolecular radical interaction. Results from fluid aqueous solutions will be used for reference and those from frozen systems will be emphasized. The implications of these findings to observations on proteins described above will be given where possible.

Electron Donation: Ionization, OH-Adduct Formation, and Hydrogen Abstraction. Ionization of amino acids or peptides either through photolysis or radiolysis leads to cation radicals, the fate of which will be influenced by the nature of the compound and the medium.

Photolytic studies of aromatic dipeptides and tripeptides in $NaClO_4$ and NaOD glasses (*50*) show that π-cation radicals of phenylalanine, tyrosine, and tryptophan can undergo different reactions depending on their disposition in the molecule and the molecular conformation. (Electrons produced upon photolysis of the perchlorate system are converted to $O^{-}\cdot$, for which corrections can be made). For PheAla and PheGly charge transfer from the $-COO^{-}$ occurs, followed by decarboxylation, as shown by reactions 8, 9, and 10:

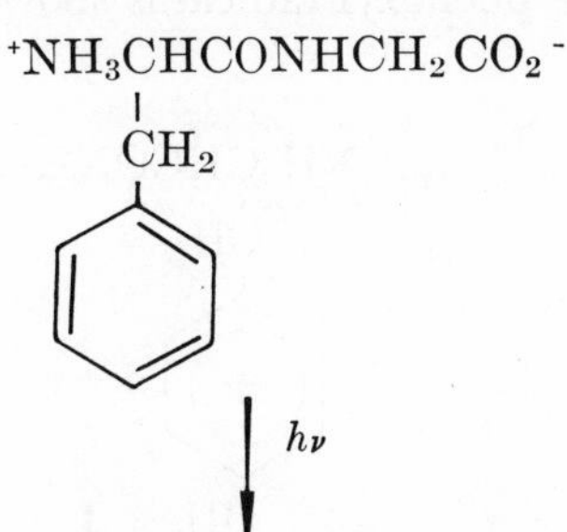

$$^{+}NH_3CH(CH_2C_6H_5)CONHCH_2CO_2^{-} \tag{8}$$

$\downarrow h\nu$

$$^{+}NH_3CH(CH_2C_6H_5^{+})CONHCH_2CO_2^{-} + e^{-} \tag{9}$$

$\downarrow$

$$^{+}NH_3CH(CH_2C_6H_5)CONHCH_2CO_2\cdot \tag{10}$$

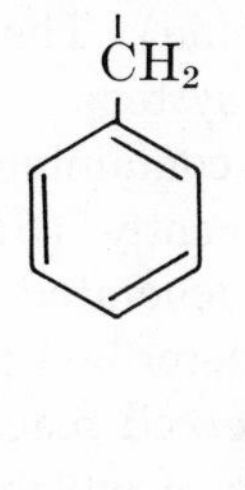

$\downarrow$

$$^{+}NH_3CH(CH_2C_6H_5)CONH\dot{C}H_2 + CO_2$$

For TyrAla, TyrGly, TrpAla, and TrpGly, radicals on the ring group are observed, along with the decarboxylated radical. Deprotonation of tyrosine leading to the phenoxyl radical is shown in reaction 11:

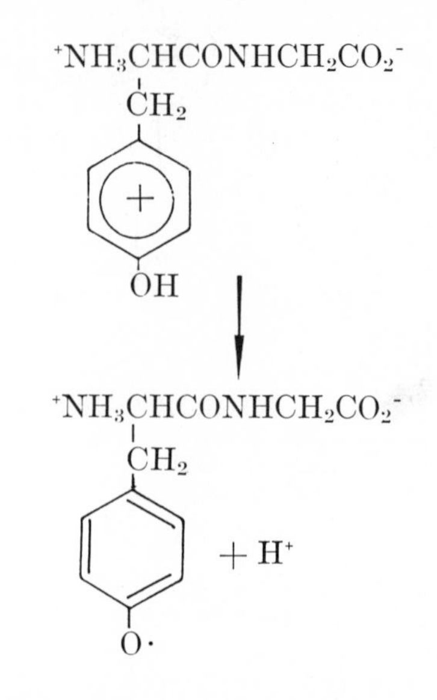

(11)

For the tripeptide, PheGlyGly, in which the cation is not favorably disposed for charge transfer, only the π-cation radical of phenylalanine is observed (this species would readily hydrolyze in water to form the OH-adduct radical of phenylalanine). These results have a direct bearing on the reactions in an irradiated system.

Radiolytic studies on ices containing peptides and *N*-acetylamino acids have been conducted recently (*51*) that provide evidence for cationic processes. (Electrons generated concurrently with the solute cations participate in reactions described below, but those formed in the ice do not contribute to the observed reactions.) The appearance of the decarboxylated species, the yield of which depends on the nature of the compound, provides this evidence. At $-196°C$, approximately 25% of the radicals observed for *N*-acetylalanine corresponds to $CH_3COND\dot{C}CH_3$, indicating that CO_2 has been lost. For the dipeptide GlyAla, the decarboxylated radical is observed to be equivalent in yield to the product from the electron reaction, indicating an efficient mechanism for decarboxylation. The findings for other dipeptides are very similar. Consistent with the photolytic results, the extent of decarboxylation is pH dependent, reflecting an influence of charge state on the mechanism.

These competitive pathways for reaction of cationic species imply that several different radicals could be formed in proteins, depending on their structure and the conditions. The evolution of CO_2 that has been observed can be understood as arising from such cationic precursors. The formation of OH-adducts of phenylalanine moieties and formation of the phenoxyl radicals from tyrosine moieties might also result from such cations and need to be investigated.

Oxidation of the amino acid moieties in irradiated aqueous systems by reaction with OH· is well established for fluid systems, but it is not likely to be encountered in frozen systems. Being a strong oxidant, the OH· reacts by electron transfer. It also adds readily to double bonds and abstracts H from C—H, N—H, and S—H bonds, but with lower reaction rate constants. A compendium of rate constants for aqueous solution has been published (*52*) and a few representative values for amino acids are shown in Table I. As discussed by Simic (*53*), the predominant sites for reaction in amino acids and peptides can be inferred from these values, which indicate that the ring groups are favored, while abstraction from the peptide backbone is less likely. Hydroxylation of the phenylalanine ring also occurs as was found for the prototype reaction with benzene (*54*). Formation of phenoxyl radical following OH· addition to tyrosine should be similar to the mechanism established for phenol (*55*) in which elimination of water occurs as is shown in reaction 12:

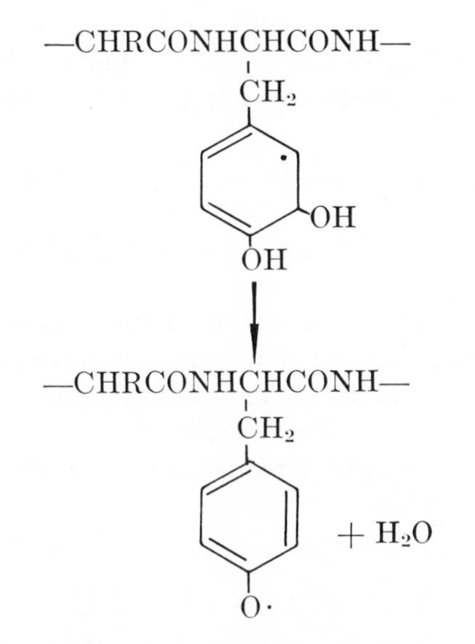

(12)

However, these adduct formation reactions are not expected to occur with much efficiency in frozen, hydrated proteins, because of the limited mobility of OH· in the rigid matrix.

Formation of α-carbon radicals either from cation decomposition or through abstraction of H by OH· is considered here because it can be construed formally as loss of an electron followed by proton transfer. If the cation deprotonates from the peptide chain by transfer to components of high proton affinity then the α-carbon radical is formed directly. Abstraction by OH· in fluid systems gives the same result.

(Abstraction of H by H· is also possible in both fluid and frozen systems. Rate constants for H· with amino acids are known (*56, 57*) and the intermediates formed have been characterized (*53*).)

Electron Acceptance: Reduction and Adduct Formation. Acceptance of electrons at specific sites on amino acids and peptides depends on their reactivities and produces different chemical consequences. Among the sites of particular importance are the terminal amino and carboxyl groups, the ring groups, the peptide carbonyl, and the sulfur bonds. Reactivities of these are reflected in the rate constants for reaction of solvated electrons with individual amino acids in aqueous solutions, as shown in Table I and as discussed by Simic (*53*). More detailed information, however, regarding the stepwise progression from attachment to specific radical formation has been obtained from low temperature studies.

Attachment to Amino Group. The electron reacts with the protonated terminal amino group in aliphatic amino acids, leading to deamination (*53, 58*). The rate constants in solution depend on pH, k decreasing as pH is increased. ESR studies on electron reaction with amino acids in neutral glasses at −196°C indicate that dissociative attachment readily occurs producing a fatty acid radical:

$$e_s^- + {}^+NH_3CHRCOO^- \rightarrow NH_3 + \cdot CHRCOO^- \qquad (13)$$

The formation of ammonia does not necessarily imply direct reaction with the $^+NH_3$ group.

Attachment to the Carboxylate Group. In basic systems where the amino group is not protonated, attachment occurs on the carboxylate ion, as evidenced from ESR studies on hydroxide glasses (*16*). At −196°C spectra are observed for the dianion radical of several amino acids. At higher temperatures, the fatty acid radicals are formed, indicating that deamination has taken place:

$$e_s^- + NH_2CHRCO_2^- \rightarrow NH_2CHRCO_2^{=}\cdot \xrightarrow{H_2O} NH_3 + \cdot CHRCO_2^- + OH^- \qquad (14)$$

The two-step sequence demonstrated for frozen systems could also occur in fluid systems, but at rates too fast to observe.

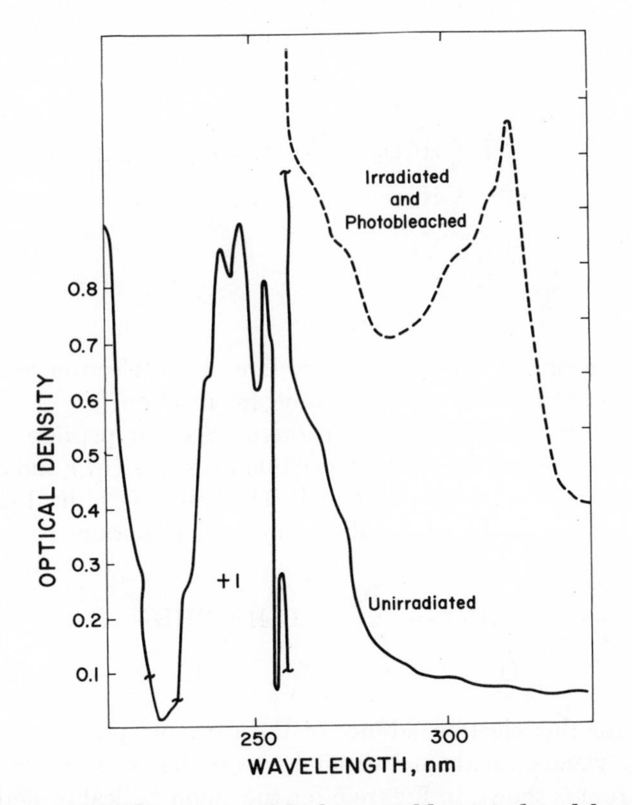

Figure 5. Optical spectrum of hydrogen adduct of phenylalanine in an ethanediol–water glass at −196°C. Glass was formed from an equimolar mixture of ethanediol and water containing 4 × 10^{-2}M phenylalanine, irradiated to approximately 4 kGy, and exposed at −196°C to visible light to bleach trapped electrons. Solid curve corresponds to phenylalanine in the unirradiated glass; dotted curve is displaced vertically for clarity and corresponds to the $^{+}NH_3CH(CH_2(C_6H_6)\cdot)CO_2^-$ radical.

ATTACHMENT TO AROMATIC AND HETEROCYCLIC GROUPS. Electrons react more rapidly with histidine, tryptophan, phenylalanine, and tyrosine than with the aliphatic amino acids, forming radicals involving the ring groups. Such electron-adduct radicals readily protonate in aqueous systems (*59*) giving the equivalent of an H-adduct radical. Similar reactions occur in frozen systems for which extensive ESR evidence has been obtained (*18*). Optical evidence for the H-adduct of phenylalanine in a water–ethanediol glass (*60*) is shown in Figure 5. Reaction 15 indicates the sequence:

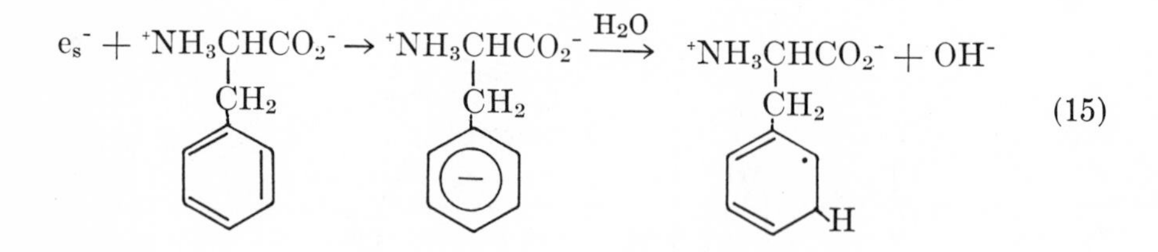

Attachment to the ring is not the exclusive fate of the electron, but it is particularly competitive with other pathways for reaction.

ATTACHMENT TO THE PEPTIDE CARBONYL. As the peptide length increases, the rate constant for electron reaction increases (*61*), indicating that reaction occurs at a peptide carbonyl. The resulting radical can be protonated depending on the pK for the following equilibrium (*62*):

$$\underset{\displaystyle O^-}{\text{—CHR}\dot{\text{C}}\text{HNR—}} \overset{H^+}{\rightleftarrows} \underset{\displaystyle OH}{\text{—CHR}\dot{\text{C}}\text{NHR—}} \qquad (16)$$

ESR evidence for this electron adduct of the carbonyl group in frozen solutions of acetylamino acids and di- and tripeptides is extensive (*50*). A typical spectrum is shown in Figure 6 for the anion radical of β-alanylglycine at $-153°C$. Depending on the peptide and pH, deamination can occur by a process involving either inter- or intramolecular electron transfer. Dissociation at other C—N bonds is also possible, leading to an amide and a "fatty acid" radical (*20, 63*); this process will be referred to as "deamidation" or "secondary deamination." As will be mentioned below, this reaction and the chemistry that follows are important for peptides and proteins.

ATTACHMENT TO SULFUR GROUPS. Of all the amino acids, the two most reactive in solution toward the electron are cystine and cysteine. Reaction with the former leads to the disulfide anion radical,

$$^+NH_3CH(COO^-)CH_2[S\overset{\bullet}{-}S]^-CH_2(COO^-)CHNH_3^+,$$

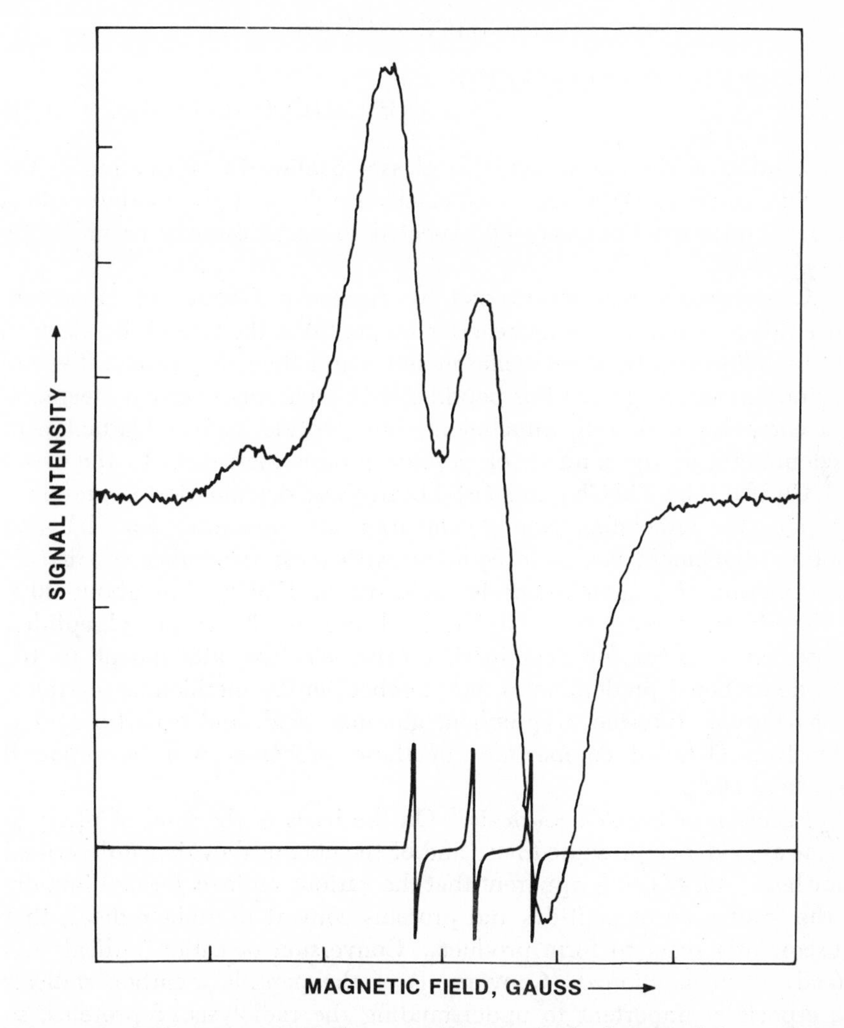

Figure 6. ESR spectrum of the peptide anion radical derived from electron attachment to β-alanylglycine at −196°C in a LiCl glass. Sample contained 5 × 10^{-2}M of β-alanylglycine and the electrons were generated photolytically (64). Spectrum was recorded at −153°C and is attributed to the radical $^{+}ND_3$-$CH_2CH_2\dot{C}O^{-}NDCH_2CO_2^{-}$.

and with the latter, to $^+NH_3CH(COO^-)CH_2\cdot$ and HS^-. Reaction with the $—CH_2SCH_3$ group in methionine also occurs and appears to produce two types of radicals:

$$e_{aq}^- + {}^+NH_3CH(COO^-)CH_2CH_2SCH_3 \rightarrow$$
$$^-SCH_3 + {}^+NH_3CH(COO^-)CH_2\dot{C}H_2 \text{ and } {}^+NH_3CH(COO^-)CH_2CH_2S^- + \cdot CH_3 \quad (17)$$

ESR studies of the amino acids in glasses confirm the formation of the disulfide anion (*20*), which is especially stable, and the methyl radical (*20*). Similar reactions may be expected to occur directly or indirectly on peptides.

COMPETITION FOR ELECTRONS BY REACTIVE GROUPS ON PEPTIDES. Since there are many sites for reaction on peptides, the fate of the electron will be influenced by the specific moieties and their disposition. Several illustrations can be given. For peptides with an aromatic group, deamination competes with ring attachment, but peptide carbonyl attachment predominates as the number of peptide groups increases. In the series (a) GlyPhe, (b) PheGly, and (c) PheGlyGly: deamination is preferred for (a), ring attachment and deamination are equivalent for (b), and peptide attachment begins to compete with these two processes in (c). Furthermore, ring attachment is exclusive in HisGly, but about 40% of the electrons react with GlyHis to deaminate it. In acetylpeptides, for which no sites for deamination exist, electron attachment to the peptide carbonyl predominates, but competition by methionine, cysteine, phenylalanine, tyrosine, tryptophan, glutamic acid, and aspartic acid is extensive. Detailed comparisons of these processes will be reported elsewhere (*64*).

Reaction of Peptide Radicals. On the basis of the kind of physicochemical evidence presented here and of the chemical evidence described elsewhere (*46, 47*), it is apparent that the various radicals formed initially in the irradiation of peptides and proteins convert to other radicals that subsequently react to form products. Conversion of cation radicals has already been mentioned. Conversions of the peptide α-carbon radicals are especially important to understanding the radiolysis of proteins, so some illustrative examples will be given. The eventual reaction of the α-carbon radicals is not well understood, but certain assumptions can be made relevant to the systems studied.

DECOMPOSITION OF THE PEPTIDE CARBONYL RADICALS. Depending on the nature of the constituent groups, these radicals can decompose by transferring the electron to the terminal amino group or by splitting off an amide. Both processes, deamination and deamidation, lead to the formation of $\cdot CHRCONH—$ radicals, deamination corresponding to chain scission. As an illustration, Figure 7 shows the spectra for D_2O ices of *N*-acetylalanine (190 mg/ml) irradiated at −196°C and then annealed (*64*). The carbonyl anion predominates in the spectrum after annealing at −153°C (to eliminate OD·) and amounts to about 65% of the radicals; the next most abundant radical is the decarboxylated species. Upon annealing to −80°C, the anion converts by deamidation to the amide and fatty acid radical:

$$CH_3\dot{C}(O^-)NDCH(CH_3)CO_2^- \xrightarrow{D^+} CH_3COND_2 + \cdot CH(CH_3)CO_2^- \quad (18)$$

The spectrum shown can be compared with that of the propionic acid radical formed independently or by the deamination of alanine. This sequence is representative of several acetylamino acids studied.

ABSTRACTION OF H FROM THE PEPTIDE BY CARBON RADICALS. The carbon radicals derived from deamination, deamidation, and decarboxylation react subsequently with the peptides to abstract hydrogen, forming the more stable α-carbon peptide radicals (*63*). This reaction can be demonstrated in irradiated ices containing acetylamino acids. Continuing with the *N*-acetylalanine example, one can also see in Figure 7 that upon further annealing to −50°C, another radical appears corresponding to reaction 19:

$$\dot{C}H(CH_3)CO_2^- + CH_3CONDCH(CH_3)CO_2^- \rightarrow CH_2(CH_3)CO_2^- + CH_3COND\dot{C}CH_3CO_2^- \quad (19)$$

The quartet spectrum shown is assigned to the radical from acetylalanine. The same type of reaction occurs with dipeptides. Consequently, a series of aliphatic dipeptides in D_2O ices were irradiated and the resultant radicals at approximately −50°C were examined to obtain detailed spectral data for comparison with proteins. Table II shows the substituents on the model peptide radical, $^+ND_3CH(R')COND\dot{C}RCO_2^-$, and the type of spectra observed. Because peptides with R = H or CH_2R'' predominate in proteins and because these would give rise to doublet splittings at low temperatures, it is understandable why the irradiated protein spectra for actomyosin, myosin, and others are characterized by a broad doublet signal. Computer simulation of the protein signal by properly combining the peptide spectra is underway.

BIMOLECULAR REACTION OF THE α-CARBON RADICALS. The eventual formation of stable covalent bonds requires that protein α-carbon radicals (whose distribution is determined by molecular conformation and the conditions) combine or disproportionate with similar or other radicals.

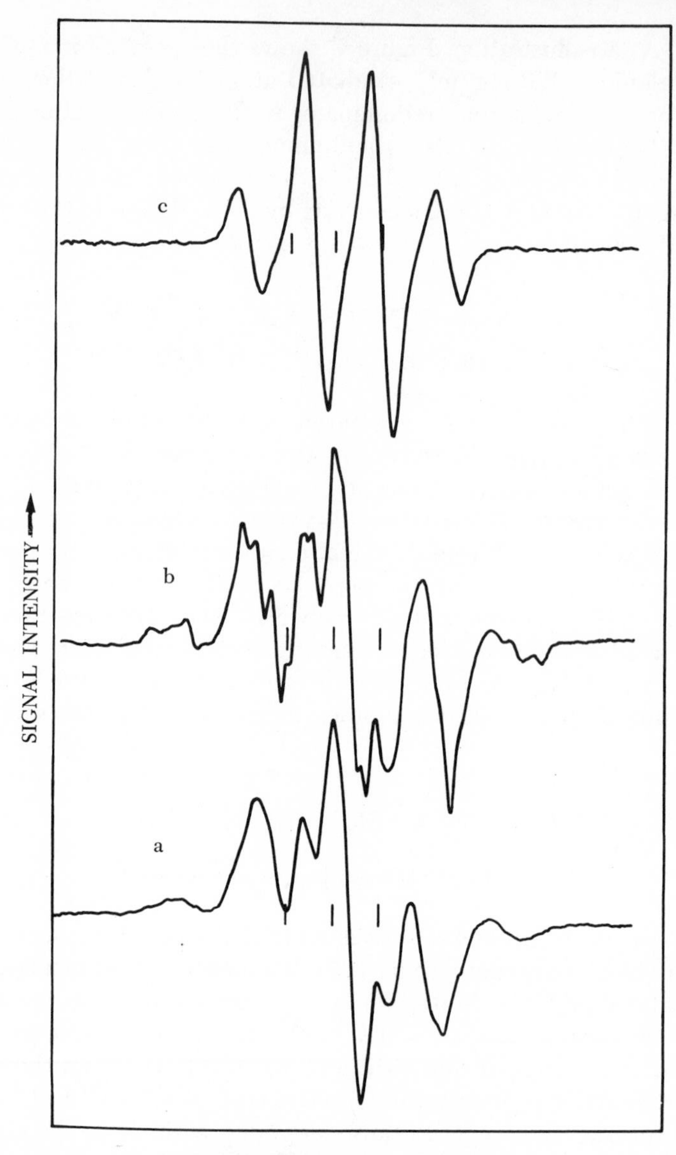

Figure 7. ESR spectra of radicals derived from N*-acetyl-*L*-alanine in an irradiated* D_2O *ice plug. Sample contained 190 mg/ml of* N*-acetylalanine and was irradiated to 5 kGy at −196°C. Spectrum a was recorded at −196°C; spectra b and c at −135°C. All spectra have markers from Fremy's salt superimposed. (a) Composite spectrum of radicals present after annealing at −153°C (approximately 65% corresponds to the anion). (b) Composite spectrum of the radicals formed upon annealing the ice plug to −80°C, of which 55% corresponds to the fatty acid radical,* $\cdot CH(CH_3)CO_2^-$. *(c) Spectrum of the peptide radical,* $CH_3COND\dot{C}(CH_3)CO_2^-$, *formed upon further annealing the ice plug to −50°C.*

Table II. ESR Spectral Features for Different $^{+}ND_3CH(R')COND\dot{C}(R)CO_2^-$ Radicals[a]

Parent Dipeptide	*R'*	*R*	*Number of Lines*[b]
Glycylglycine	H	H	2
Alanylglycine	CH_3	H	2
Glycylalanine	H	CH_3	4
Alanylalanine	CH_3	CH_3	4
Glycylglutamic acid	H	$CH_2CH_2CO_2^-$	2
Glutlmylglutamic acid	$CH_2CH_2CO_2^-$	$CH_2CH_2CO_2^-$	2
Glycylaspartic acid	H	$CH_2CO_2^-$	2
Glycylmethionine	H	$(CH_2)_2SCH_3$	2
Glycylserine	H	CH_2OH	2
Lysyllysine	$(CH_2)_4ND_3^+$	$(CH_2)_4ND_3^+$	2

[a] Species stable after γ-irradiating approximately 200 mg/ml of the corresponding dipeptide in ice plugs at −196°C to a dose of about 5 kGy and annealing to −50°C.

[b] Hyperfine splittings are found to be 18–20 gauss for all the radicals observed.

Their large size and relative immobility would render such reactions slow in fluid media and improbable in viscous or rigid media. Reaction in dry systems and in frozen systems would involve either small mobile radicals diffusing to the larger radicals or free radical sites on the large molecules being in proximity to each other. Fragment radicals such as $\dot{C}H_3$, $\dot{C}HRCO_2^-$, or $\dot{S}CH_3$ could account for some of the decay. For large, globular polypeptides or proteins, the radicals on folded-over chains could combine if close enough. For the long, fibrous molecules, the radicals on neighboring chains might be able to react if sufficient bending occurs to place these sites in an appropriate disposition. Since myosin has two chains entwined about each other, some cross reaction might be expected. If the proteins are desiccated there should be some unraveling in the structure, making reaction even less likely. Significantly more kinetic information is needed before these final steps in the radiolytically initiated sequence of reactions are understood.

Summary

Many of the reactions that occur in the specific systems described are common to most irradiated proteins. These reactions are summarized in the generalized scheme given below. No attempt is made to show all possible pathways for reaction or to explain all of the observations noted for proteins. For the sake of clarity, the protein structure is idealized and schematically represented by /\/\/\/\/\/, which is equivalent to (20) (details in the structure are included only for specific cases):

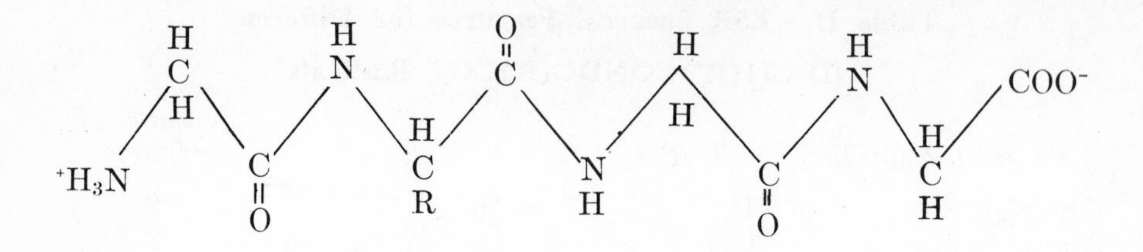

$$H_2O \xrightarrow{\gamma} e_s^-, OH\cdot, H^+, H\cdot \qquad (21)$$

/\/\/\ $\xrightarrow{\gamma}$ ionization → /\/\/\⁺ $+ e^- (\rightarrow e_s^-)$ (22a)

excitation → (/\/\/\)* (22b)

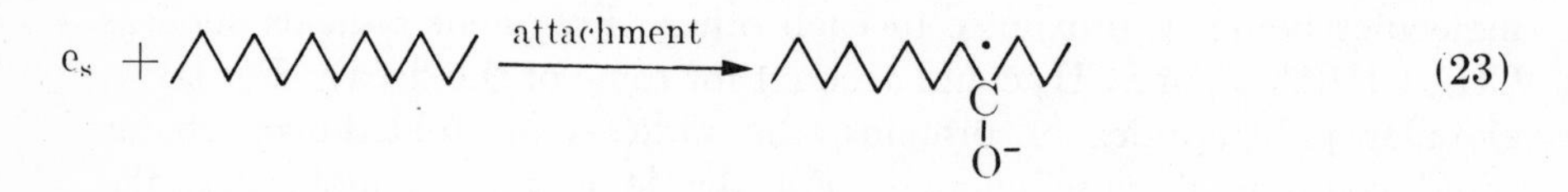

$OH\cdot +$ /\/\/\ — abstraction → [structure] $+ H_2O$ (24a)

addition–elimination → [structure] $+ H_2O$ (24b)

O·

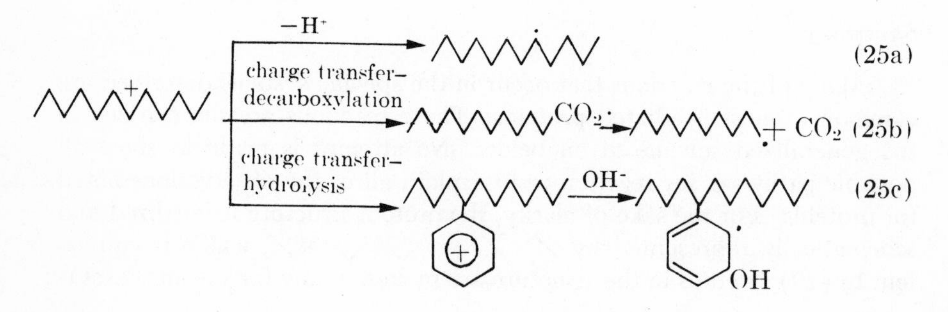

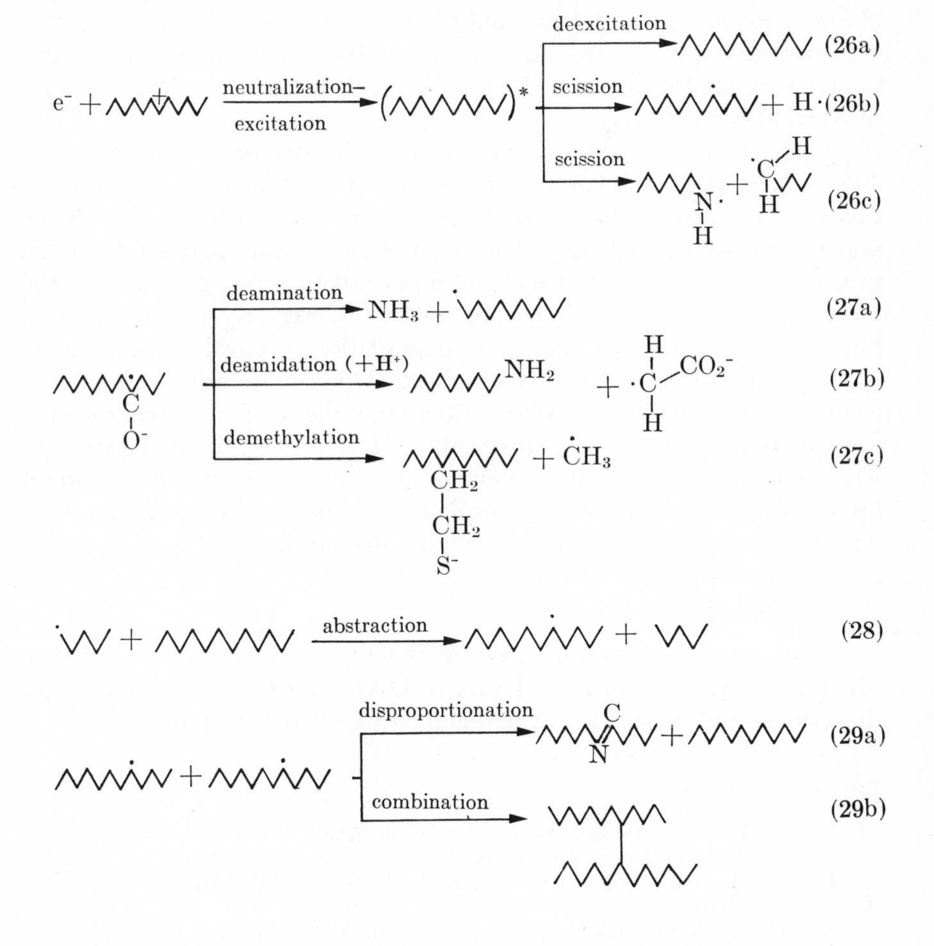

As is implied in this scheme, there will be certain modifications in the protein resulting from irradiation that depend on specific conditions. The indirect effects of reactions of primary water radical, as well as those resulting from unimpeded diffusion of secondary radicals, will be minimized or eliminated in dry systems or frozen aqueous systems. Formation of the peptide radical by long chain, fatty acid-type radicals abstracting hydrogen could occur in fluid, concentrated solutions, but would be difficult in rigid media, being limited to either neighboring molecules or proximal parts of the same molecule. Effects on globular and fibrous proteins will differ for this reason. Final reaction of the large peptide radicals, in turn, would be affected by the constraints of the medium on their flexing and diffusional motions. Such reactions in frozen, hydrated, fibrous proteins might involve only neighboring radicals brought close enough together by flexing and torsional modes of motion. In dry systems, the interaction of these would be more difficult because of a less favorable disposition of the chains.

Since the proteins in food preserved by irradiation to approximately 40 kGy at −40°C are hydrated and fixed in a rigid medium, the observation (*6*) that these proteins are minimally affected is consistent with the implications of this scheme. The major free radicals derived from the protein are of the peptide backbone type, and they do not persist in the thawed product. These radicals apparently undergo reconstitutive recombination reactions, so no significant degradation or covalent aggregation of the long molecular chain is observed. Other free radicals from side chain groups represent a small contribution to the total, and consequently, none of the amino acids is discernibly affected. The limited extent to which changes occur in the proteins, as well as in the equally important lipid components (*65, 66, 67*), explains the high quality of the low-temperature irradiated meat, fish, and poultry products.

Acknowledgment

The authors are grateful to Dr. Michael G. Simic, Dr. James J. Shieh, Mr. John E. Walker, and Mr. James B. D'Arcy for helpful discussions of this subject and for making available data in advance of publication.

Literature Cited

1. Spinks, J. W. T.; Woods, R. J. "An Introduction to Radiation Chemistry"; 2nd ed.; John Wiley: New York, 1976.
2. Hart, E. J.; Anbar, M. "The Hydrated Electron"; Wiley-Interscience: New York, 1970.
3. Anbar, M.; Bambenek, M.; Ross, A. B. "Selected Specific Rates of Reactions of Transients from Water in Aqueous Solution. Hydrated Electron," *Natl. Bur. Stand. (U.S.) Rep.* **1973**, NSRD–NBS 43.
4. Taub, I. A.; Kaprielian, R. A. *Abstr. First Int. Cong. Eng. Food, Aug 9–13, 1976, Boston, MA;* p 82.
5. Taub, I. A.; Kaprielian, R. A.; Halliday, J. W. *IAEA Proc. Int. Symp. Food Preserv. Irradiation, Nov. 21–25, 1977* **1978,** *1,* 371.
6. Taub, I. A.; Robbins, F. M.; Simic, M. G.; Walker, J. E.; Wierbicki, E. *Food Tech.* **1979,** *33,* 184.
7. Rustgi, S.; Joshi, A; Friedberg, F.; Riesz, P. *Int. J. Radiat. Biol. and Relat. Stud. Phys. Chem. Med.* **1977,** *32,* 533.
8. Matheson, M. S.; Dorfman, L. M. "Pulse Radiolysis"; MIT: Cambridge, 1969.
9. Taub, I. A.; Hurwitz, P. A.; Tocci, J. *Abstr. Fifth Int. Cong. of Radiat. Res., 1974,* Seattle, WA, p 196.

10. Trifunac, A. D., Johnson, K. W.; Clifft, B. E.; Lowers, R. H. *Chem. Phys. Lett.* **1975**, *35*, 566.
11. Fessenden, R. W. *J. Chem. Phys.* **1973**, *58*, 2489.
12. Smaller, B.; Avery, E. C.; Remko, J. R. *J. Chem. Phys.* **1971**, *55*, 2414.
13. Asmus, K. D. In "Fast Processes in Radiation Chemistry," Adams, G. E., Fielden, E. M., Michael, B. D., Eds.; John Wiley: New York, 1975; p 40.
14. Verberne, J. B.; Warman, J. M.; deHaas, M. P.; Hammel, A.; Prinsen, L. *Nature* **1978**, *272*, 343.
15. Eiben, K.; Fessenden, R. W. *J. Phys. Chem.* **1971**, *75*, 1186.
16. Sevilla, M. D. *J. Phys. Chem.* **1970**, *74*, 2096.
17. Ibid. **1976**, *80*, 1898.
18. Van Paemel, C.; Frumin, H.; Brooks, V. L.; Failor, R.; Sevilla, M. D. *J. Phys. Chem.* **1975**, *79*, 839.
19. Sevilla, M. D.; Brooks, V. L. *J. Phys. Chem.* **1973**, *77*, 2954.
20. Sevilla, M. D. *J. Phys. Chem.* **1970**, *74*, 3366.
21. Henriksen, T.; Nelo, T. B.; Sexebol, G. In "Free Radicals in Biology," Pryor, W. A., Ed.; Academic: New York, 1976; Vol. II, p 213.
22. Satterlee, L. D.; Wilhelm, M. S.; Barnhart, H. M. *J. Food Sci.* **1971**, *36*, 549.
23. Giddings, G. G.; Markakis, P. *J. Food Sci.* **1972**, *37*, 361.
24. Tappel, A. L. *Food Res.* **1956**, *21*, 650.
25. Simic, M. G.; Taub, I. A.; Rosenkrans, R. L. *J. Food Proc. Preserv.*, in press.
26. Simic, M. G.; Taub, I. A. *Biophys. J.* **1978**, *24*, 285.
27. Shieh, J. J.; Hoffman, M. Z.; Simic, M. G.; Taub, I. A. "Abstracts of Papers," 174th National Meeting, ACS, Aug. 28–Sept. 2, 1977, Chicago, IL; BIOL 86.
28. Shieh, J. J.; Sellers, R. M.; Hoffman, M. Z.; Taub, I. A., *Proc. Assoc. Radiat. Res. (Great Britain), June 3–5, 1979, Wrexham, Clwyd, Wales,* in press.
29. Shieh, J. J., unpublished data.
30. Mee, L. K.; Adelstein, S. J.; Stein, G. *Radiat. Res.* **1972**, *52*, 588.
31. Adams, G. E.; Bisby, R. H.; Cundall, R. B.; Redpath, J. L.; Willson, R. L. *Radiat. Res.* **1972**, *49*, 290.
32. Delincée, H.; Radola, B. J. *Int. J. Radiat. Biol. and Relat. Stud. Phys. Chem. Med.* **1975**, *28*, 565.
33. Friedberg, F. *Radiat. Res.* **1969**, *38*, 34.
34. Fluke, D. J. *Radiat. Res.* **1966**, *28*, 677.
35. Riesz, P.; White, F. H., Jr. *Radiat. Res.* **1970**, *44*, 24.
36. Coelho, R. *Enzymol. Aspects Food Irradiat., Proc. Panel, 1968,* **1969**, 1.
37. Taub, I. A.; Halliday, J. W.; Holmes, L. G.; Walker, J. E.; Robbins, F. M. *Proc. Army Sci. Conf., June 6–9, 1976, West Point, NY;* Vol. III.
38. Halliday, J. W.; Taub, I. A. *J. Food Proc. Preserv.*, in press.
39. Henriksen, T.; Sanner, T.; Pihl, A. *Radiat. Res.* **1963**, *18*, 147.
40. Gordy, W.; Shields, H. *Radiat. Res.* **1958**, *9*, 611.
41. Halliday, J. W.; Taub, I. A., unpublished data.
42. Taub, I. A.; Halliday, J. W.; Holmes, L. G.; Walker, J. E.; Robbins, F. M., unpublished data.
43. Stein, G.; Tomkiewicz, M. *Radiat. Res.* **1970**, *43*, 25.
44. Bachman, S.; Galant, S.; Gasyna, Z.; Witkowski, S. *IAEA Proc., Panel Improvement Food Quality, 1973,* **1974**, 77.
45. Hayden, G. A.; Rogers, S. C.; Friedberg, F. *Arch. Biochem. Biophys.* **1966**, *113*, 247.
46. Garrison, W. *Radiat. Res. Rev.* **1972**, *3*, 305.
47. Friedberg, F. *Radiat. Res. Rev.* **1969**, *2*, 131.
48. Meyers, L. S., Jr., In "Radiation Chemistry of Macromolecules," Dole, M., Ed.; Academic: New York, 1970; Vol. II, p. 323.

49. Alexander, P.; Lett, J. T. In "Comprehensive Biochemistry," Florkin, M.; Stotz, E. H.; Eds.; Elsevier: New York, 1971; Vol. 27, p 267.
50. Sevilla, M. D.; D'Arcy, J. B. *J. Phys. Chem.* **1978**, *82*, 338.
51. Sevilla, M. D.; D'Arcy, J. B., unpublished data.
52. Dorfman, L. M.; Adam, G. E. "Reactivity of Hydroxyl Radical in Aqueous Solution," *Nat. Bur. Stand. (U.S.), Rep.* **1973**, NSRD–NBS 46.
53. Simic, M. G. *J. Agric. Food Chem.* **1978**, *26*, 6.
54. Dorfman, L. M.; Taub, I. A.; Buhler, R. E. *J. Chem. Phys.* **1962**, *36*, 3051.
55. Land, E. S.; Ebert, M. *Trans. Faraday Soc.* **1967**, *63*, 1181.
56. Neta, P. *Chem. Rev.* **1972**, *72*, 533.
57. Volker, W. A.; Kuntz, R. R. *J. Phys. Chem.* **1968**, *72*, 3394.
58. Weeks, B. M.; Cole, S.; Garrison, W. M. *J. Phys. Chem.* **1965**, *69*, 4631.
59. Hayon, E.; Simic, M. *Acc. Chem. Res.* **1974**, *7*, 114.
60. Taub, I. A.; Heinmets, F., unpublished data.
61. Simic, M.; Hayon, E. *Radiat. Res.* **1971**, *48*, 244.
62. Rao, P. S.; Hayon, E. *J. Phys. Chem.* **1974**, *78*, 1193.
63. Garrison, W. M.; Jayko, M. E.; Rodgers, M. A. J.; Sokol, H. M.; Bennett–Corniea. *Adv. Chem. Ser.* **1968**, *81*, 384.
64. Sevilla, M. D.; D'Arcy, J. B., unpublished data.
65. Nawar, W. W. *J. Agric. Food Chem.* **1978**, *26*, 21.
66. Merritt, C. M.; Angelini, P.; Graham, R. A. *J. Agric. Food Chem.* **1978**, *26*, 29.
67. Simic, M. G.; Merritt, C. M.; Taub, I. A. In "Fatty Acids," *AOCS Monograph;* Pryde, E. H., Ed.; in press.

RECEIVED June 16, 1978.

7

The Role of Proteins in the Freezing Injury and Resistance of Biological Material

J. LEVITT

Carnegie Institution, Department of Plant Biology, 290 Panama Street, Stanford, CA 94305

The first protein hypothesis explained freezing injury by a precipitation of proteins due to the freeze-induced increase in salt concentration. This concept is opposed by the absence of a reduction in the activity of most enzymes in freeze-killed plants. Many results point to membranes, and more specifically to their ATPases as the initial locus of freezing injury. Due to their role as ion pumps, their inactivation by freezing is the probable cause of the well-known efflux of ions from freeze-injured tissues. Freezing tolerance of plants is apparently associated with protein changes in both quality and quantity, and also with the accumulation of protective substances that prevent ATPase inactivation.

FREEZING INJURY (*1*)

Effect of Freezing on Animals and Plants

The conversion of water to ice in temperate climates poses a serious survival problem for all organisms exposed to this change. All higher animals and most lower animals are killed when frozen in nature. Those that survive freezing temperatures do so primarily by freezing avoidance —the ability to remain unfrozen at environmental freezing temperatures. In contrast, all overwintering plants of temperate climates (the higher as well as the lower ones) possess freezing tolerance. They survive at the temperatures of their natural environment. The biologist, and even the philosopher of earlier civilizations, has long been fascinated by this extraordinary adaptation of plants—their ability to freeze solid without

0-8412-0484-5/79/33-180-141$5.00/0

suffering any injury. This adaptability is of more practical interest because of man's transfer of economically important plants to environments to which they are not adapted. Many agricultural, forest, and ornamental plants are killed each year by freezing. How does the freezing of tissues kill them and how do resistant plants prevent this freezing injury?

Physical Explanations of Freezing Injury

Many hypotheses have been proposed to explain freezing injury. Because freezing is a physical phenomenon, these hypotheses have been mainly physical in nature. The early theorists viewed freezing injury as a simple rupture of cells due to the expansion of water on freezing, analogous with the bursting of water pipes in winter. Nearly 150 years ago this theory was proved to be wrong by a very simple experiment—the observation of frozen cells under the microscope. Hundreds of species of plants were observed without finding any rupture of frozen and thawed cells. Under artificial conditions of freezing it has since been found that some thin-walled cells can be ruptured by freezing, but this has never been observed in cells frozen in nature. The reason for the absence of rupture was also discovered a century and a half ago. When frozen naturally, plant tissues freeze extracellularly, in the air-filled intercellular spaces, at the expense of water that gradually diffuses out of the cells to the extracellular loci of ice. The cells are dehydrated in the proccess and collapse to a greater or less degree, depending on the temperature to which they are frozen and their water contents before freezing (Figures 1 and 2). Freezing of plants in winter is, therefore, a dehydration process, analogous to the evaporative dehydration of plant cells during a summer drought. Freezing injury and drought injury must, therefore, be basically similar. As a result, bacteria, yeasts, and other microorganisms that can be air dried without injury can also survive extracellular freezing without injury.

Under some artificial conditions, however, plants freeze intracellularly due to the much more rapid freezing (e.g., a cooling rate of 1°/min) which does not permit sufficient time for the water to diffuse out of the cells to the extracellular loci of ice. Even then, cell rupture does not normally occur because of the elastic extensibility of the cell walls. The intracellular ice crystals may, however, pierce the cell membranes. This leaves holes on thawing, allow the cell contents to diffuse out, resulting in death of the cell.

Under natural conditions of slow freezing, the cell collapse that accompanies extracellular freezing subjects the cells to another kind of physical stress, as pointed out 50 years ago by Iljin. Instead of expanding, as occurs during intracellular freezing, slow freezing results in collapse of the cell because of freeze dehydration, and this subjects the cell surface to tension. In its simplest sense, this tension may be thought of as producing cracks or crevices in the cell surface, or more specifically in the cell membrane. Such postulated cracks and crevices would not be seen microscopically because they presumably occur in bimolecular membranes. Thus, if they do occur they must take place at the molecular level and chemical reactions may be involved. However, a stronger argument can be made for involvement of chemical reactions in freezing injury. This concerns the long known fact that the hardening process, during which the plant increases its freezing resistance, involves many chemical changes (see below).

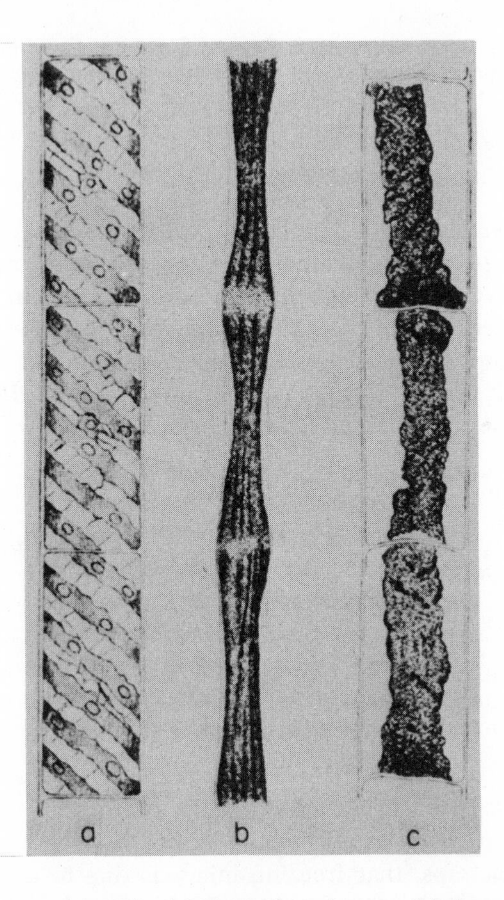

Figure 1. Spirogyra (×300). (a) Normal, unfrozen, (b) frozen extracellularly showing cell collapse without any ice inside the cell, (c) thawed. From Molisch, 1897.

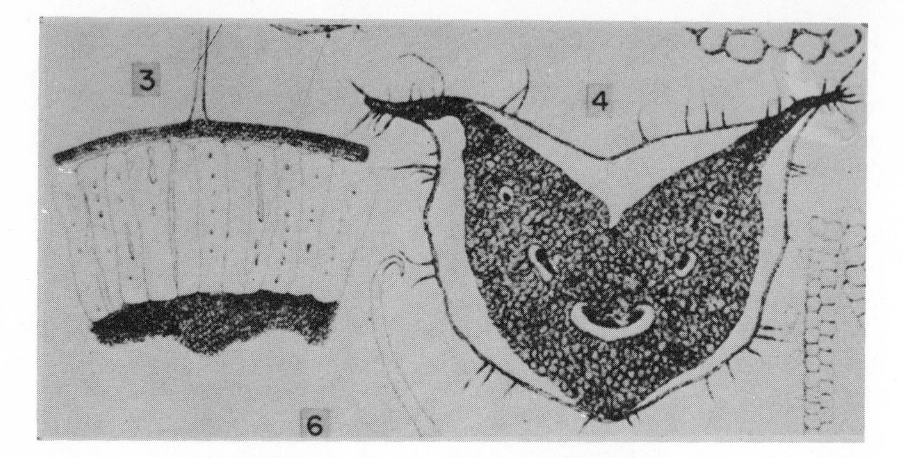

Figure 2. (Left) Large ice masses formed between the much smaller cells. (Right) Contraction of tissue caused by ice masses formed beneath the epidermis. From Prillieux, 1869.

Effect of Freezing on Cell Proteins

Protein Precipitation. The living part of a cell is the protoplasm, which consists of proteins, lipids, nucleic acids, a source of energy (usually carbohydrates), metabolic intermediates, certain inorganic ions, and cofactors necessary for the enzyme activity of many proteins. Because the proteins comprise the largest component of protoplasm (Table I), it is logical to look first at them as the seat of cell injury. It is perhaps for

Table I. Components of Protoplasm

Substance	*Percent of Fresh Weight*	*Approximate Relative Number of Molecules*
Water	85–90	18,000
Proteins	7–10	1
Fatty substances	1–2	10
Other organic substances	1–1.5	20
Inorganic ions	1–1.5	100

this reason that the first chemical theory of freezing injury involved proteins. Some 70 years ago, it was suggested by Gorke, on the basis of freezing of plant saps, that freezing injury is due to a precipitation of proteins, that occurs primarily because of a freeze-induced increase in the concentration of cell salts. This salt precipitation theory of Gorke's was reasonable at the time, and a similar hypothesis was proposed independently some 50 years later to explain the hemolysis of red blood cells by freezing. It is now called the "salt concentration hypothesis" and is still the most popular explanation among medical cryobiologists. More recent attempts to support Gorke's hypothesis have, however, led to failure (see below).

Effect of Freezing on Enzyme Activity. If proteins play any role in freezing injury, an effect of freezing on their enzyme activity should be detectible. That freezing temperatures, per se, do not inactivate enzymes in general has been repeatedly shown by investigators in the new field of cryoenzymology—the measurement of enzyme activity at temperatures below the freezing point of water (*2*). These activities, however, are generally measured in the absence of ice, which is avoided by the addition of antifreezes. More pertinent evidence is easily derived from the fact that freezing is used as a routine method for preserving enzyme activity in tissues killed by freezing (*3, 4*). The activities of the enzymes so obtained are quite comparable to those of enzymes obtained by conventional extraction (*3*). The presence of soluble, fully active enzymes in freeze-killed cells is, of course, the strongest evidence against theories of freezing injury that involve irreversible precipitation of proteins. Nevertheless, some enzymes are inactivated by freezing (*5, 6, 7*), and this may involve changes in enzyme conformation (*8*).

In all the above cases, freezing was performed rapidly and, therefore, undoubtedly involved intracellular freezing. If, however, even this more severe kind of freezing fails to inactivate most enzymes, certainly the less severe extracellular freezing would also fail to do so. This fact has been corroborated by the evidence of undiminished enzyme activity in naturally frozen and killed plants. Therefore, freezing injury cannot be due to a general inactivation of enzymes. Most enzymes, in fact, must be excluded from the cause of freezing injury. One or a few enzymes, however, may conceivably be involved.

Effects of Low Temperature on Isolated Enzymes

In order to explain the possible role of proteins in freezing injury, evidence must be borrowed from the effects of low temperatures on isolated enzymes and other model systems. There are two major, disruptive effects of low but nonfreezing temperatures (0–5°C) on protein structure. (1) An effect on the quaternary structure: the spontaneous splitting of high MW multimers into two or more small MW monomers (*9*). (2) An effect on the tertiary structure: the unfolding (denaturation) of native proteins. In both cases, the change is due to a weakening of intramolecular, hydrophobic bonds (*10*) (see Figure 3). In both cases, the change is reversible:

$$\mathrm{N} \underset{\text{above } 5^{\circ}\mathrm{C}}{\overset{0\text{–}5^{\circ}\mathrm{C}}{\rightleftharpoons}} \mathrm{D}$$

where N is the native and D the denatured protein.

This denaturation may inactivate the enzyme, but the unfolding is reversible, and the protein refolds and regains its enzymatic activity on rewarming. It is, perhaps, for this reason that cooling below 0°C in the absence of freezing (undercooling) is not injurious to plants that are nevertheless killed by freezing at the same temperature.

The denaturation may, however, be converted to an irreversible state by intermolecular aggregation (A):

$$\mathrm{D} \rightarrow \mathrm{A}$$

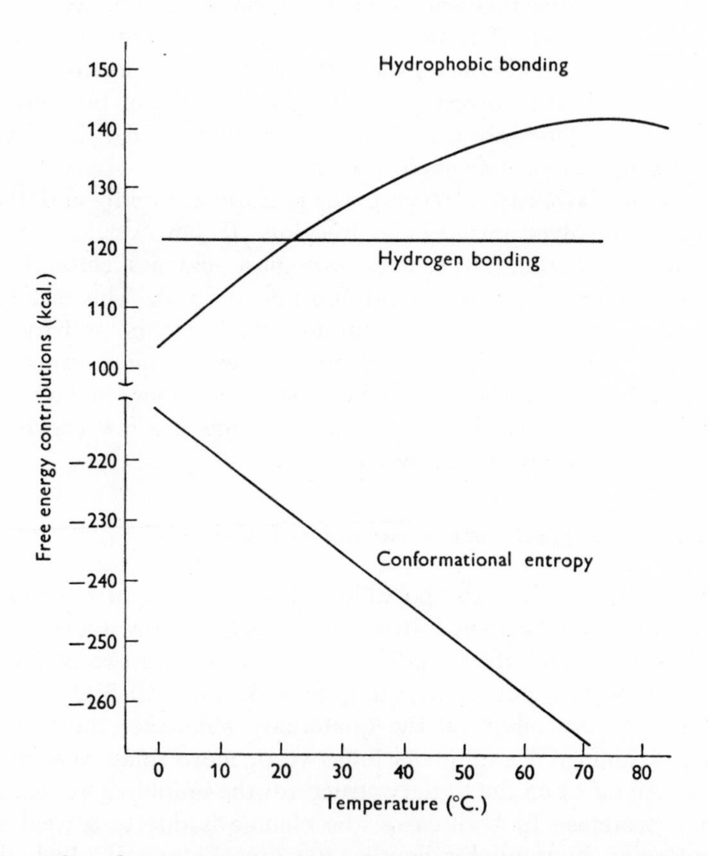

Figure 3. The relation of hydrophobic bonding, hydrogen bonding, and conformational entropy to temperature. From Brandts (1967).

This aggregation may involve any of the known kinds of intermolecular protein bonds: H—, hydrophobic, electrostatic, or covalent (S—S). Of these four, by far the strongest and the least likely to be reversed simply by raising the temperature is the S—S bond.

The Sulfhydryl–Disulfide Hypothesis of Freezing Injury

On the basis of experimental results indicating a conversion of protein SH to SS groups during freezing, the SH hypothesis of freezing injury was proposed (*11*). According to this concept, freeze-dehydration partially removes the free water that separates proteins in the unfrozen living cell. This permits close enough contact between SH groups of adjacent protein molecules for intermolecular SS formation, resulting in aggregation and denaturation (Figure 4). This order is contrary to the order expressed in

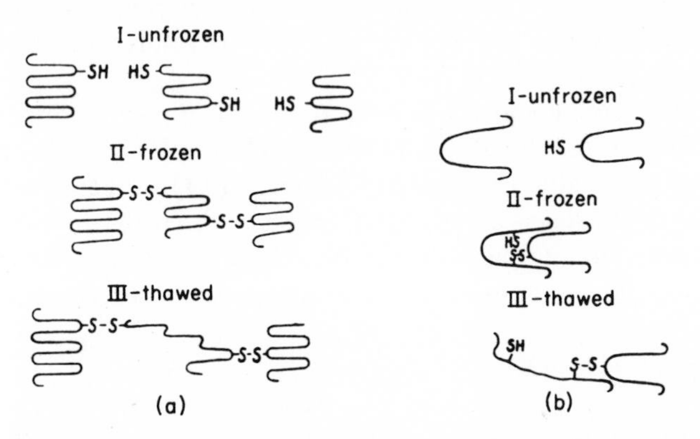

Figure 4. Postulated mechanisms of protein unfolding due to intermolecular SS formation during freezing. From Levitt 1962.

the above two equations. The SH hypothesis was, therefore, later modified so that unfolding (N⇌D) occurred first (thus unmasking the SH groups that were normally protected in the folded, native proteins). Freeze-dehydration then permitted the unfolded molecules to approach close enough to permit irreversible aggregation via intermolecular bonding.

More evidence is available for the second (D→A) than for the first (N⇌D) occurrence. As mentioned above, it has long been known that freezing of plant homogenates may precipitate proteins. This precipitation is due to protein aggregation and has been explained by the freeze-induced increase in concentration of salt or of H^+ ions. Whether or not a similar precipitation occurs within the freeze-damaged cell has not been established. By use of disc gel electrophoresis, however, it has been shown that both extracellular freezing (*12*) and evaporative dehydration of plant tissues (*13*) lead to decreased migration of the faster moving, low molecular weight proteins, and increased migration of the slower moving, higher molecular weight proteins. This evidence is in agreement with the notion that some soluble, low molecular weight proteins may be induced to aggregate on freezing.

In the case of two model systems, freeze-induced aggregation is the result of intermolecular S—S bonding. In one of these cases, an aqueous gel of a denatured and thiolated gelatin (Thiogel) was surface frozen, leading to partial freeze-dehydration of the gel. This resulted in intermolecular bonding due to oxidation of SH groups to intermolecular SS groups (Figure 5). In the second case, the SS protein, bovine serum albumin (BSA), was not precipitated or aggregated if a solution of the native protein was frozen. If, however, its intramolecular SS groups were first reduced to SH groups, denaturing the protein, freeze-dehydration resulted in intermolecular SS formation, aggregation, and precipitation. In both model systems, the aggregates could be resolubilized by reducing the SS groups to SH groups.

Negative evidence also has been obtained. Freeze-induced inactivation of ATPase in the thylakoids of spinach chloroplasts has been observed without any detectable increase in SS groups (*14*). This is not surprising because, in the absence of soluble proteins (which the investigators had washed away), no aggregation is likely. A more recent investigation has demonstrated that evaporative dehydration does, indeed, lead to some

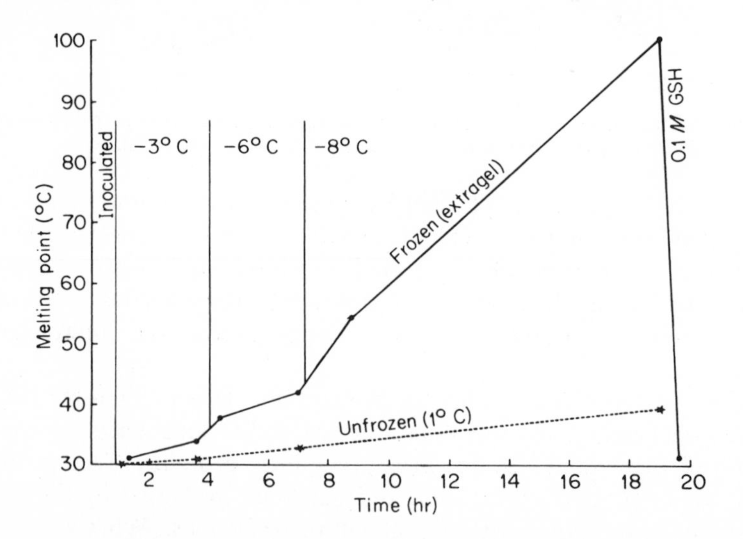

Figure 5. Acceleration of intermolecular SS formation (shows by rise in melting point) in Thiogel on freezing. From Levitt, 1965. GSH = reduced glutathione. "Extragel" freezing is external to the gel, on the upper and lower surfaces.

conversion of SH to SS groups in the soluble proteins of some plants, although it was not possible to account for the change quantitatively. This was due, at least partly, to a synthesis of new proteins during the dehydration period (*13*), which changed the proportions of SH and SS groups in the extracted mixture of soluble proteins. Furthermore, the aggregation of soluble proteins was reversed by the addition of mercapto-ethanol, a SH compound. Aggregation decreased the number of isolated protein bands from eight to three and mercaptoethanol increased the number back to eight, although the regenerated proteins were not identical to the original ones.

No direct investigation of the proposed first step—unfolding of the proteins at low temperatures—has appeared. Indirect evidence in favor of this occurrence is the reported decrease in hydrophobicity of proteins during hardening (see below). In opposition, however, is the fact that evaporative dehydration can apparently lead to the second step (aggregation) without the first. Conditions of this experiment seem to eliminate the first step ($N \underset{\text{above } 5^\circ}{\overset{0\text{–}5^\circ}{\rightleftarrows}} D$) because the evaporative dehydration was conducted at room temperature. However, it is conceivable that unfolding of the protein (the first step) may occur in the absence of low temperature. In the case of cell dehydration, whether by extracellular freezing or evaporation, Iljin has shown that cell collapse is accompanied by a tension at the cell surface (see above). This tension must be transmitted to the surface proteins, and may be expected to induce at least partial or localized unfolding. The unfolding would be aided by the dehydration- or freeze-induced increase in salt concentration, because this could break some of the electrostatic bonds that help maintain the native conformations of the proteins.

Specific Enzyme(s) or Protein(s) Involved in Freezing Injury

In 1912, Maximov proposed that the locus of freezing injury is the outer cell membrane, the plasmalemma. Many later investigators have also suggested that membrane damage is the seat of injury. If this is correct, it would point to membrane proteins as the important ones in freezing injury. If the damage is due to aggregation, as suggested above, then the soluble cytoplasmic proteins must also be involved, because these are the only ones that can be brought in contact with the membrane proteins by freeze-dehydration.

In agreement with the membrane concept, Heber (*15*) investigated a number of enzymes and found that none of the soluble ones were inactivated by freezing. The only one affected by freezing was the

ATPase of chloroplast membranes (the thylakoids). Recent indirect evidence supports these results, although ATPase of the plasmalemma appears to be the most likely key enzyme. This conclusion resulted from a reinvestigation of a long-known effect of freezing. In 1932, Dexter showed that the degree of freezing injury could be evaluated quantitatively by measuring the efflux of ions from frozen tissue immediately after thawing. This was accomplished simply by measuring the conductivity of the solution leached from the thawed tissue. The greater the injury, the more rapid the efflux of ions, and therefore the higher the conductivity of the leachate. It was generally assumed that the ion efflux was from dead cells, and that the conductivity was, therefore, proportional to the percent dead cells on thawing. Direct observation (*16*) has now revealed that conductivity increases profoundly even in the absence of dead cells, and if injury is not too severe, all the cells are able to recover completely. The conductivity increase was, therefore, due to damage to the cell membrane that was sufficient to increase efflux of ions, but was insufficient to cause death of the cells. Membrane damage is, therefore, apparently the first step in freezing injury. If this membrane is sufficiently severe, the cells subsequently die. This could, of course, be interpreted to mean that the injury involved membrane lipids rather than membrane proteins. Direct measurements eliminated this interpretation because permeability of the cells to water did not change despite a marked increase in ion efflux. The membrane lipids were, therefore, functionally intact and presumably the passive efflux of ions also remained unaltered. These two results can be explained only by a decrease in the active uptake of ions, leading to an increased net efflux. This active uptake is dependent on ATPases, which act as ion pumps. Freezing injury is therefore presumably initiated by inactivation of ATPases in the plasmalemma.

In the absence of direct evidence, the major support for this concept arises because it is in agreement with, and can explain, all existing information on the nature of freezing injury. (1) It agrees with the conclusion that freezing injury is initiated in the membrane, specifically the plasma membrane. (2) The ATPase of plasma membranes also has been found to be freezing labile (*17*). (3) ATPases are SH proteins and may require some SH groups for activity. (4) ATPases exist in the membrane where the greatest tension occurs when the cell collapses during dehydrative, extracellular freezing. Therefore, they are the proteins most in danger of being unfolded by the cell tension, thereby exposing previously masked, internal SH groups. Freeze-dehydration would simultaneously induce close contact with soluble proteins in the cytoplasm (which may also be unfolded) leading to possible intermolecular SS bonding and therefore, irreversible aggregation. This would inactivate the ATPase irreversibly, due to its inability to refold on thawing. (5) Siminovitch and his coworkers (*18*) have found that the phospholipid content of cells is the factor most closely related to freezing resistance (see below). ATPases are, in fact, lipoproteins, and depend on the attached lipid for their ion pump activity. Furthermore, if cell collapse is sufficiently severe, the ATPase, which presumably has become anchored to the cytoplasm by aggregation with a soluble protein, may conceivably pop out of the membrane, either during cell contraction on freezing, or during its re-expansion on thawing. This could produce additional damage over and above inactivation of ATPase, because removal of the enzyme from the membrane may also remove the attached lipid, leading to permeable holes in the membrane.

The proposed series of changes leading to freezing injury is presented in Figure 6.

Extracellular freezing
↓
Dehydrative cell collapse
↓
Tension at cell surface plus
increase in salt concentration
↓
Reversible unfolding ($N \rightleftharpoons D$)
of plasma membrane ATPase
↓
Exposure of previously masked SH
and other potentially bonding groups
↓
Simultaneous close approach of
soluble proteins to membrane, due
to removal of free H_2O
↓
Aggregation ($D \rightarrow A$) of ATPase with
soluble (unfolded) proteins, via SS
and other bonds
↓
Inactivation of ATPase
↓
Net efflux of K^+ and other ions
↓
Cell injury

Figure 6. Proposed scheme of changes leading to injury during extracellular freezing

FREEZING RESISTANCE

Degrees and Kinds of Resistance

Freezing resistance in plants ranges from those that are killed by the first touch of frost (no resistance), to those that can be plunged into liquid N_2 or even exposed to temperatures within a fraction of a degree of absolute zero without suffering injury. Those that are killed by slight freezing are called tender and those that survive freezing are called hardy. Plants are classified as slightly hardy when they survive roughly −5°C or a little below, moderately hardy when they survive temperatures of −10°C to −20°C, very hardy when they survive −20°C to −30°C, and extremely hardy when they survive below −30°C.

There are different kinds of freezing resistance and some of these are in no way related to the cell's proteins. For instance, the living cells in the wood of trees from temperate climates can survive freezing temperatures as low as −40°C by remaining in an undercooled state. This kind of resistance is apparently dependent of cell wall properties, and proteins play no role. On the other hand, the undercooling of fish is definitely due to a specific glycoprotein, or more correctly a glycopolypeptide that prevents the freezing of water at temperatures slightly below (about −1 to −2°C) the true freezing point of the solution (*19*).

There are several kinds of freezing resistance, such as supercooling, accumulation of antifreeze, and so on, in which proteins can play no direct role. A direct role has been indicated only in the case of freezing tolerance. This, however, is by far the most important kind of freezing resistance in plants. Without it, no plant could survive the presence of ice.

The Hardening Process

One feature of hardy higher plants is highly unusual. Even the most resistant plants, which can be plunged into liquid N_2 without injury, possess this ability only in midwinter. The same plants in late spring or early summer are killed if frozen at about −5°C. Between late summer and early winter, these plants undergo a hardening process which increases their freezing resistance from the above-mentioned low of −5°C to the full resistance of midwinter. Most of this hardening may occur within about a two-week period in the fall. Many investigators have followed the physico-chemical changes that occur during this hardening process, and therefore, there is a large body of information that should help to explain the role of proteins in freezing resistance. If proteins are, indeed, the locus of freezing injury, changes associated with hardening should reveal whether freezing resistance is due to (1) changes in the proteins as such or (2) an accumulation of substances capable of protecting proteins against freeze-induced denaturation and aggregation.

Protein Changes during Hardening

Many investigators have shown that soluble proteins accumulate within plant cells during the hardening period. This occurrence does not, of course, prove a cause and effect relation. There is, in fact, an accumulation of many substances, especially soluble carbohydrates, at the same time. This happens because the relatively low temperatures (0 to 5°C) to which plants are exposed during the hardening period, and the concurrent accumulation of growth inhibitors, retards or stops growth, while metabolism (including photosynthesis in evergreens and winter annuals) continues at a decreased but still significant rate. The result is that certain compounds are synthesized more rapidly than they are used for growth, resulting in an accumulation of these compounds (Figure 7).

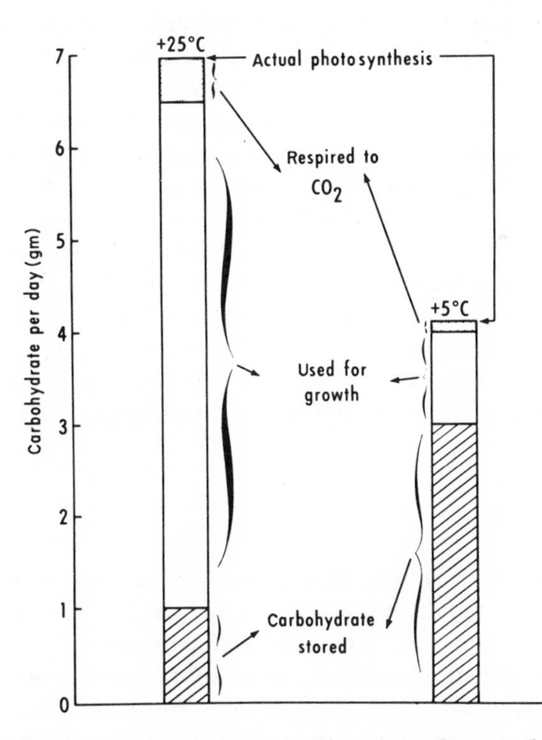

Figure 7. Schematic representation of changes in photosynthesis, growth, and carbohydrate storage, due to low (hardening) temperatures. From Levitt 1972.

The aforementioned accumulation of proteins may, perhaps, occur simply to permit the burst of growth that plants exhibit in early spring before rapid photosynthesis is possible. It may, therefore have little or nothing to do with freezing tolerance. In support of this conclusion, Siminovitch and his coworkers showed that when protein accumulation is prevented by artificial methods, considerable hardening (enough to survive freezing at −20°C) still occurs. Maximum hardening, however, does not occur in the absence of protein accumulation. Perhaps this is so because there is a need for rapid metabolism at low, hardening temperatures, and that this need is met by an increase in enzyme concentration. It has been suggested that the increase in enzyme concentration that occurs during hardening permits metabolic processes to take place almost as rapidly at 0 to 5°C (the hardening temperature) as at the normal growth temperature.

In order to play a more direct role in freezing tolerance, the accumulation of proteins must be accompanied by (1) a change in the structure of proteins or enzymes, and/or (2) synthesis of new proteins or enzymes. Some evidence for both of these changes has been produced. With regard to point 1, the proportions of amino acids in a major protein (RNDPCase) of cabbage leaves change during hardening, with the most hydrophobic ones decreasing and the hydrophilic ones increasing (*20*). Evaporative dehydration, which also hardens the plant, has the same effect (*13*). This change would result in the replacement of weaker hydrophobic interactions by stronger hydrophilic interactions (mostly intramolecular hydrogen bonds) and this would oppose the tendency to unfold ($N \rightleftharpoons D$) at the low temperatures.

With regard to point 2, synthesis of two new proteins has been observed during the hardening of wheat seedlings (*21*). The amino acid composition of these proteins again indicates a higher degree of hydrophilicity than in the original proteins. Several other investigators also have observed the formation of new proteins during hardening (e.g., *22*). These results all support the theory that proteins play an active role in freezing tolerance.

Protective Substances

It has long been known that sugars (sometimes sugar alcohols) accumulate during the hardening of many plants. Due to their high solubility, they can protect the plant colligatively—by decreasing the amount of freeze-dehydration at any given subfreezing temperature—and thereby lessen cell shrinkage during slow freezing. This mechanism presumably imparts only moderate hardiness, since nearly all the free water freezes at about −20°C, even in the presence of high concentrations of sugars. However, due to the high solids content (soluble plus insoluble) of the hardy cell, the maximum possible cell shrinkage is greatly reduced and so is the tension on the membrane.

The existence of substances capable of protecting proteins more specifically against freeze-induced damage will be discussed by Heber. One kind of protection directly pertinent to the SH hypothesis must, however, be mentioned. A high reductive capacity of the plant could conceivably lessen formation of intermolecular SS bonds in proteins during freezing. This reductive capacity would require that the plant continue photosynthetic accumulation of reducing substances at hardening temperatures and at the same time decrease carbohydrate synthesis so that this enhanced reducing capacity will not be lost (Figure 8).

On the basis of the above concepts, the mechanism of freezing tolerance may depend on several factors. (1) Membrane ATPase and soluble proteins may be protected colligatively (by sugars, etc.) so freeze-induced unfolding, aggregation and inactivation does not occur. (2) Soluble proteins may undergo decreased hydrophobicity and a corresponding increase in the strength of intramolecular hydrophilic bonds, thereby stabilizing a folded conformation. In the absence of unfolding, reactive SH and other groups would remain masked, thus preventing the soluble proteins from aggregating membrane ATPase. (3) A reductive environ-

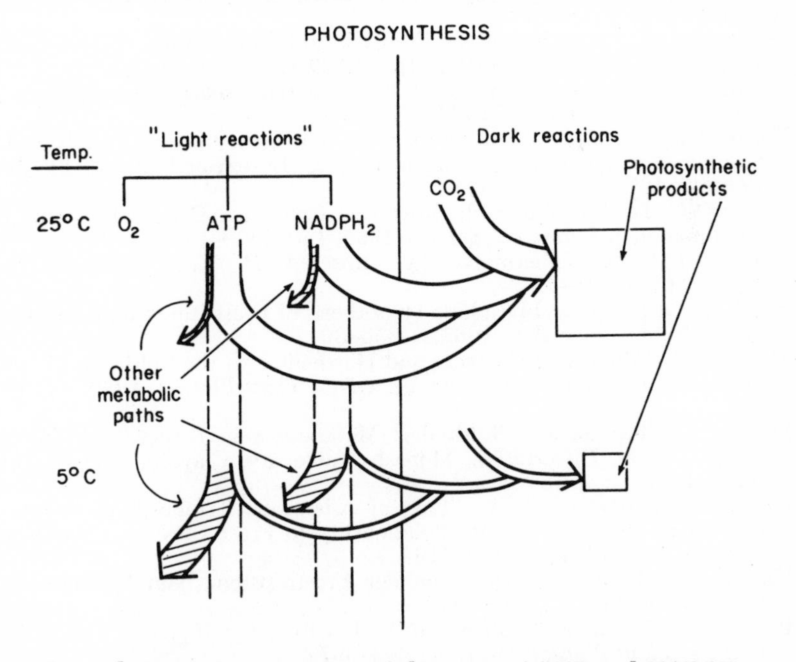

Figure 8. Schematic representation of diversion of ATP and NADPH to non-photosynthetic pathways at 5°C. From Levitt, 1972.

ment would prevent conversion of protein SH groups to intermolecular SS bonds and might therefore prevent irreversible aggregation. (4) Even if a considerable fraction of the membrane ATPase is inactivated during freezing, the large reserve of proteins and phospholipids may serve to quickly repair the membrane after thawing.

Which, if any, of the above possible mechanisms are responsible for the freezing resistance of specific plants connot be accurately determined without further study.

Literature Cited

1. Levitt, J. "Responses of Plants to Environmental Stresses"; Academic: New York, 1972.
2. Fink, A. L. "Cryoenzymology: The Use of Subzero Temperatures and Fluid Solutions in the Study of Enzyme Mechanisms," *J. Theor. Biol.* **1976,** *61,* 419–445.
3. Rhodes, D.; Stewart, G. R. "A Procedure for the In Vivo Determination of Enzyme Activity in Higher Plant Tissue," *Planta* **1974,** *118,* 133–144.
4. Judel, G. K. "Influence of pH and Temperature on the Activity of Phenol Oxidase and Peroxidase from Sunflower Plants," *Biochem. Physiol. Pflanz.* **1975,** *167,* 243–252.
5. Anderson, J. O.; Nath, J. "The Effects of Freeze Preservation on Some Pollen Enzymes. I. Freeze-Thaw Stresses," *Cryobiology* **1975** *12,* 160–168.
6. Darbyshire, B. "The Results of Freezing and Dehydration of Horseradish Peroxidase," *Cryobiology* **1975,** *12,* 276–281.
7. Whittam, J. H.; Rosano, H. L. "Effects of the Freeze–Thaw Process on α-Amylase," *Cryobiology* **1973,** *10,* 240–243.
8. Yu, N.-T.; Jo, B. H. "Comparison of Protein Structure in Crystals and in Solution by Laser-Raman Scattering: I. Lysozyme," *Arch. Biochem. Biophys.* **1973,** *156,* 469–474.
9. Markert, C. L. "Lactate Dehydrogenase Isozymes: Dissociation and Recombination of Subunits," *Science* **1965,** *140,* 1329–1330.
10. Brandts, J. F. In "Thermobiology," Rose, A. H., Ed.; Academic: New York, 1967; pp 25–72.
11. Levitt, J. "A Sulfhydryl-Disulfide Hypothesis of Frost Injury and Resistance in Plants," *J. Theor. Biol.* **1962,** *3,* 355–391.
12. Morton, W. "Effects of Freezing and Hardening on the Sulfhydryl Groups of Protein Fractions from cabbage leaves, *Plant Physiol.* **1969,** *44,* 168–172.
13. Daniel, V. "Protein and Sulfhydryl Metabolism in Desiccation Tolerant Plants," Ph.D. Dissertation, Monash University, Clayton, Victoria, Australia, 1977.
14. Heber, U.; Santarius, K. A. "Loss of Adenosine Triphosphate Synthesis Caused by Freezing and Its Relationship to Frost Hardiness Problems," *Plant Physiol.* **1964,** *39,* 712–719.
15. Ullrich, H.; Heber, U. "Ursachen der Frostresistenz bein Winterweizen. IV," *Planta* **1961,** *57,* 370–390.
16. Palta, J. P.; Levitt, J.; Stadelmann, E. J. "Freezing Injury in Onion Bulb Cells. 1" *Plant Physiol,* **1977,** *60,* 393–397.
17. Steponkus, P. L.; Wiest, S. C. "Freezing Injury of Plant Plasma Membranes," *Cryobiology* **1973,** *10,* 532.

18. Siminovitch, D.; Sings, J.; de la Roche, I. A. "Studies on Membranes in Plant Cells Resistant to Extreme Freezing. I," *Cryobiology* **1975**, *12*, 144–153.
19. De Vries, A. I. "Cold Resistance in Fishes in Relation to Protective Glycoproteins," *Cryobiology* **1970**, *6*, 585.
20. Shomer-Ilan, A.; Waisel, Y. "Cold Hardiness of Plants: Correlation with Changes in Electrophoretic Mobility, Composition of Amino Acids and Average Hydrophobicity of Fraction-I-Protein," *Physiol. Plant.* **1975**, *34*, 90–96.
21. Rochat, E.; Therrien, H. P. "Étude des Protéines des Blés Résistants, Kharkov, et Sensible, Selkirk, au Cours de l'Endurcissement au Froid. I. Protéines Solubles," *Can. J. Bot.* **1975**, *53*, 2411–2416.
22. Volger, H. G.; Heber, U. "Cryoprotective Leaf Proteins," *Biochem. Biophys. Acta,* **1975**, *412*, 335–349.

RECEIVED June 16, 1978.

18. [illegible] [illegible] Freezing [illegible] 19[illegible]
19. [illegible] "Cold acclimation [illegible] Relation to Frost [illegible]" Cryobiology 1970, 6, 586.
20. [illegible] "Cold Hardiness of Plants [illegible]" [illegible]
21. [illegible] [illegible]
22. [illegible] 19[illegible]

Received [illegible] 1978

8

Membrane Damage and Protection During Freezing

U. HEBER, H. VOLGER, V. OVERBECK, and K. A. SANTARIUS

Institute of Botany, University of Dusseldorf, 4000 Dusseldorf, West Germany

Freezing can damage biomembranes either mechanically or by solute action. Damage is described using mainly thylakoid membrane vesicles as examples. After freezing in suitable media, light-dependent ion gradient formation across thylakoids no longer occurs, photophosphorylation is inactivated, membrane semipermeability is lost and electron transport is either stimulated or decreased. Membranes aggregate and polar membrane proteins dissociate. A main factor in freezing damage is the action of potentially membrane-toxic solutes which are concentrated during ice formation. Membrane toxicity of salts follows a Hofmeister power series. The mechanism of membrane inactivation is discussed.

Freezing damage can be prevented by adding cryoprotectants. Protection occurs by unspecific colligative solute action and by specific membrane stabilization. Sugars, sugar alcohols, and specific proteins are physiological cryoprotectants. Membranes must be protected on both sides against freeze-inactivation.

In 1954, Lovelock reported that hemolysis of red blood cells, which occurred during freezing in physiological saline, was caused by membrane damage (*1*). Not the ice crystals formed below the freezing point, but the accumulation of salts accompanying ice formation was thought to be responsible for injury (*2*). Several years later, photophosphorylation of thylakoid membranes isolated from chloroplasts of leaf cells was found to be inactivated during freezing in the presence of certain solutes (*3, 4, 5*). Again, mechanical membrane damage, though it occurs during

0-8412-0484-5/79/33-180-159$7.75/0

eutectic freezing (*6, 7*), was not responsible for loss of activity. Mitochondrial oxygen uptake and phosphorylation, both membrane-bound processes, were also decreased by freezing (*8, 9*).

Since many other cell constituents are not altered by freezing, the inactivation of membrane activities suggests that damage to biomembranes is responsible for the sensitivity of cells and organisms to freezing (*10–15*). In this contribution, we will describe conditions which lead to membrane damage. Damage will be characterized using the well investigated thylakoid membrane as an example.

Many organisms are known to acquire freezing resistance under suitable environmental conditions. Obviously, freezing does not damage their membranes. In view of potential applications, particularly in agriculture, medicine, and food preservation, it is important to know how membrane damage can be avoided. We will briefly consider mechanisms responsible for membrane protection during freezing.

Freezing

In the most simple case, when a suspension of membranes in a solution containing only one solute is frozen, water is converted to ice and the solute concentration rises in the unfrozen part of the system which also contains the membranes. Figure 1 shows the relation between freezing temperature and the composition of a binary system consisting of water and a salt. When the temperature is lowered below the freezing point and the initial salt concentration is not too high, ice crystals will separate from the solution. The mole fraction of the salt in the unfrozen part of the system increases as the temperature is further decreased. When the eutectic temperature T_E is reached, the system solidifies to a maximum extent. Addition of a third component to a binary mixture shifts the eutectic temperature to lower values. Further shifting is produced by the addition of more components. In complex biological systems, where high viscosity may also retard crystallization of solutes, eutectic freezing is rarely observed. When it occurs, extensive membrane damage is produced, apparently because the membranes are mechanically disrupted by the mass of ice and solute crystals formed (*6, 7*). Eutectic freezing of sea water in which sea urchin eggs were suspended killed all cells (*16*).

When freezing is slow, ice will form outside of membrane-enclosed spaces, and the membranes become exposed to increased solute concentrations. Vesicular membranes respond osmotically. Increased solute concentrations also produce other effects. As will be shown, the concentration of certain solutes is an important factor in the alteration of biomembranes during freezing, and this can lead to loss of membrane

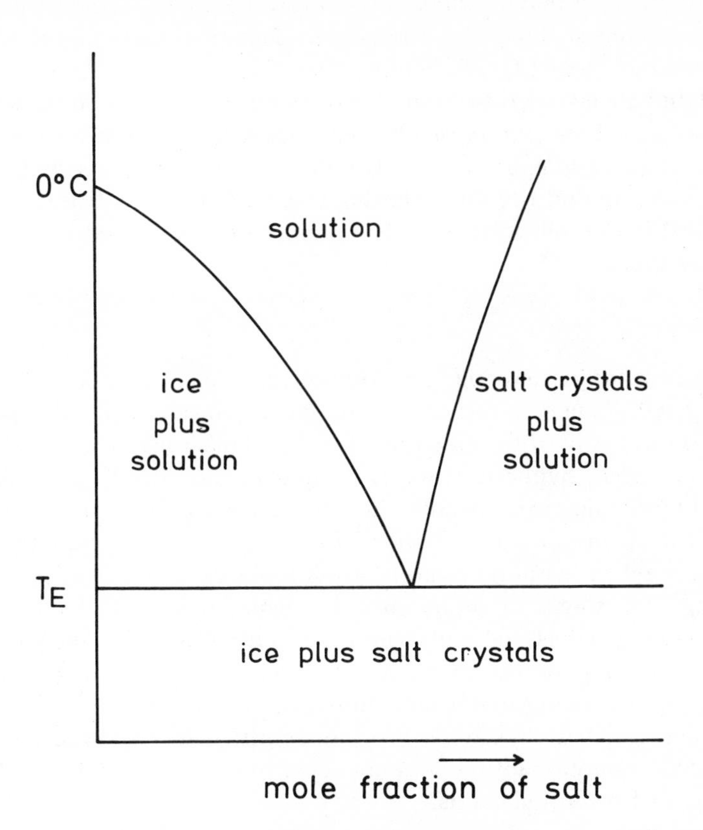

Figure 1. Relation between temperature and composition of a single salt solution. T_E *is eutectic temperature*

function. Solute damage should, at first sight, be expected to increase as the sub-freezing temperature is decreased because the solute concentration in the unfrozen part of a system increases with decreasing temperature. However, the changes in membrane structure which lead to loss of membrane function are temperature-dependent processes. Thus, while solute stress increases as the temperature is lowered, the rate of membrane inactivation may actually be faster at higher temperatures than at lower temperatures (*6*).

When freezing is too fast to allow transport of intravesicular water to extravesicular ice loci, ice formation takes place inside membrane-enclosed spaces. This appears to be mechanically disruptive. The rate of freezing is therefore an important parameter of freezing injury. Slow freezing prolongs exposure of membranes to high solute concentrations in a temperature range where solute damage may proceed rapidly (*6*), and fast freezing may produce mechanical membrane damage in cellular systems by intracellular ice formation. The two-stage theory of injury proposed by Mazur (*17, 18, 19*) describes these relations. During freezing of isolated thylakoid membranes suspended in an isotonic or somewhat hypotonic medium, intravesicular ice formation and mechanical damage do not occur as long as eutectic freezing is avoided by a suitable choice of the medium and the final freezing temperature. The freezing damage described in the following work is therefore usually caused by high solute concentrations.

Thawing

During rapid thawing, membranes may be flooded by water from melting ice. Where this occurs, osmotic expansion will take place. If water transport by diffusion is too slow to abolish major gradients in the water potential, hypnotic stress is created in the vicinity of melting ice, while hypertonic stress persists in other parts of the system. As intact biological membranes are closed and osmotically active, osmotic rupture can occur when hypnotic osmotic stress replaces solute stress. As during freezing, two stages of injury may be distinguished. Especially during slow thawing, effects of high solute concentrations may increase injury in a critical temperature range. Alternatively, osmotic damage may occur during rapid thawing under nonequilibrium conditions.

Since isolated thylakoids are less sensitive to hypnotic stress than intact cells, osmotic rupture does not occur to any significant extent during thawing of frozen thylakoids.

Solute Effects During Freezing

Inorganic Salts. Red blood cells suspended in 0.15 *M* NaCl first shrink during freezing, while the osmolarity of the unfrozen part of the suspending medium increases. When it reaches about 0.8 *M*, NaCl leaks in, K^+ is lost, and the cells start to hemolyze on thawing. Freezing is not necessary for the effect to occur, as a gradual increase in the NaCl concentration to 0.8 *M* or more at 0°C also causes osmotic cell shrinkage and then hemolysis (*2*). However, with intact cells it is rather difficult to decide which effect produces injury.

In the chloroplasts of higher plants, closed chlorophyll-containing membranes called thylakoids function to convert light energy into chemical energy. They can be isolated and, without too much loss of biochemical activity, largely freed from soluble cell constituents by cautiously washing them in hypotonic solutions or even water (*20, 21*). During washing, they swell but do not lose their osmotic properties. When resuspended in media containing nonpenetrating solutes, they shrink. Isolated thylakoids constitute a much simpler test system than intact cells for measuring the susceptibility of biomembranes to freezing.

When thylakoids are frozen in a dilute salt medium (5 m*M* NaCl), their capability to form ATP and ADP and phosphate in the light is much decreased (*22, 23*). If the NaCl concentration is increased before freezing, loss of activity during freezing is even more pronounced. Inorganic salts such as NaCl increase freezing damage. In the absence of freezing, low concentrations of NaCl do not inactivate photophosphorylation. When a cryoprotective compound is present with NaCl during freezing, the osmolar ratios of these solutes determine whether membrane function is preserved or lost.

Figure 2 shows inactivation of thylakoids after 3 hours of storage at −6 or −12°C as a function of the NaCl concentration in the suspending medium before freezing. Controls were kept for 3 hours at 0°C. 100 m*M*

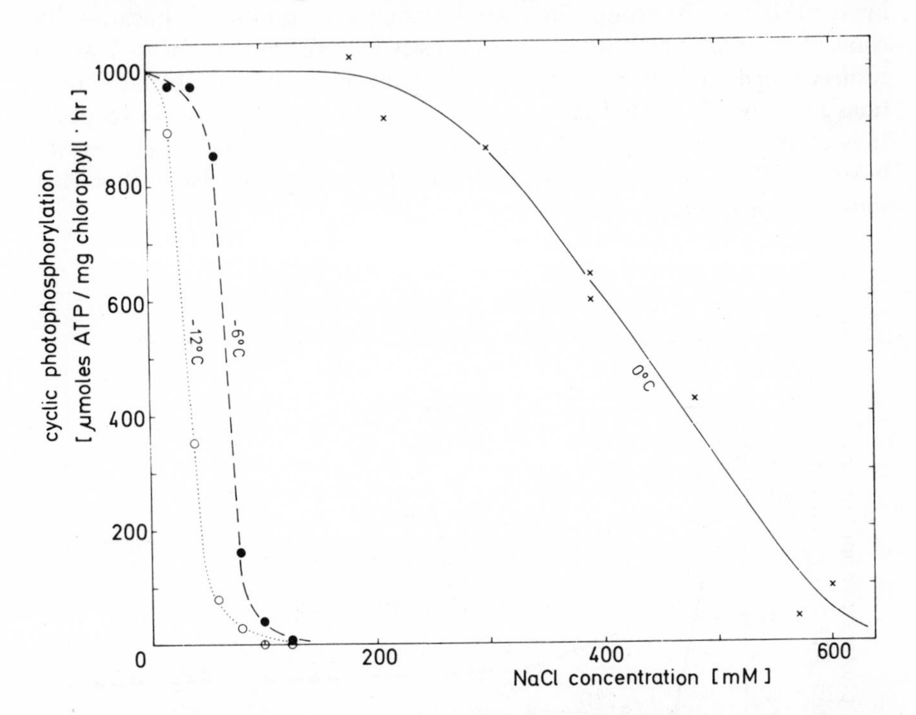

Figure 2. Effect of temperature on inactivation of thylakoids in the presence of NaCl. Washed thylakoids were suspended in a solution containing 100 mM *sucrose and NaCl and were kept for 3 hours at 0°C, −6°C and −12°C. Freezing and thawing were fairly rapid and final temperatures were reached within less than 2 minutes. Sucrose served as cryoprotectant and was added to prevent freeze-inactivation of the membranes in the presence of low salt concentrations. After thawing, the activity of cyclic photophosphorylation was measured. Experimental conditions have been described previously* (5, 20, 21).

sucrose, which is a cryoprotectant in the thylakoid system, was also added to prevent membrane inactivation caused by low concentrations of added NaCl during freezing. At $-12°C$, an initial concentration of about 50 m*M* NaCl in the suspending medium was sufficient to overcome cryoprotection by sucrose and produce significant membrane damage. Naturally, the NaCl concentration in equilibrium with ice was much higher than the initial concentration. At $-6°C$, about 100 m*M* NaCl in the suspending medium were required to produce extensive damage. Still higher concentrations were needed to damage the membranes at 0°C (*6, 24*). Thus, biomembranes as widely differing as blood cell membranes and photosynthetic membranes share a similar sensitivity to NaCl. The NaCl-photosynthetic membranes share a similar sensitivity to NaCl. The NaCl-induced loss of membrane function during freezing or at 0°C is irreversible.

In Figure 2, membrane damage increases as the temperature is lowered below freezing. Increased damage is produced because the concentration of NaCl increases progressively (above that shown) as the subfreezing temperature is decreased. However, rates of chemical reactions are directly related to temperature. Solute damage is no exception. It is therefore not surprising that, as the temperature is further lowered below $-12°C$, the extent of solute damage may decrease rather than show a further increase (*6*).

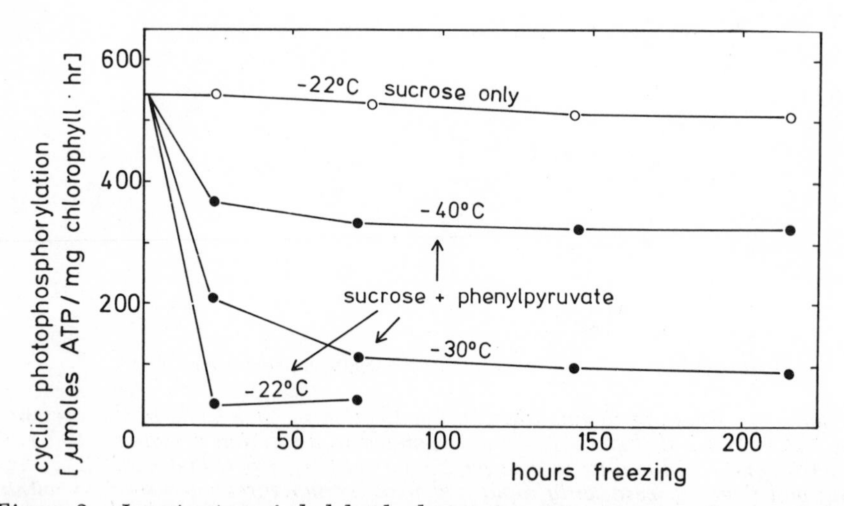

Figure 3. Inactivation of thylakoids during freezing at various low temperatures as a function of time. Washed thylakoids were suspended in a solution containing 50 mM *sucrose as a cryoprotectant and 20* mM *sodium phenylpyruvate as a cryotoxic solute. The suspensions were rapidly frozen and thawed. After thawing, photophosphorylation was determined. For experimental conditions, see notes in legend for Fig. 2*

Figure 3 shows inactivation of thylakoids suspended in a solution containing a cryotoxic solute (20 m*M* sodium phenylpyruvate) and a cryoprotectant (50 m*M* sucrose). Controls contained sucrose only. There was no significant inactivation of thylakoid function of the controls during 9 days of storage at −22°C. In the presence of phenylpyruvate, damage was essentially complete after 24 hours at −22°C. Damage was decreased as the temperature was decreased to −30°C or to −40°C. In these experiments, partial crystallization of sodium phenylpyruvate may have occurred during freezing and this may have reduced the effective concentration of the cryotoxic agent, particularly at the lower temperatures. This effect may have contributed to the striking differences in membrane damage seen at the different temperatures.

CATIONS. During freezing, thylakoids suffer more damage in the presence of LiCl than in the presence of isosmolar concentrations of NaCl (Figure 4). For different alkali metal chlorides, damage decreased

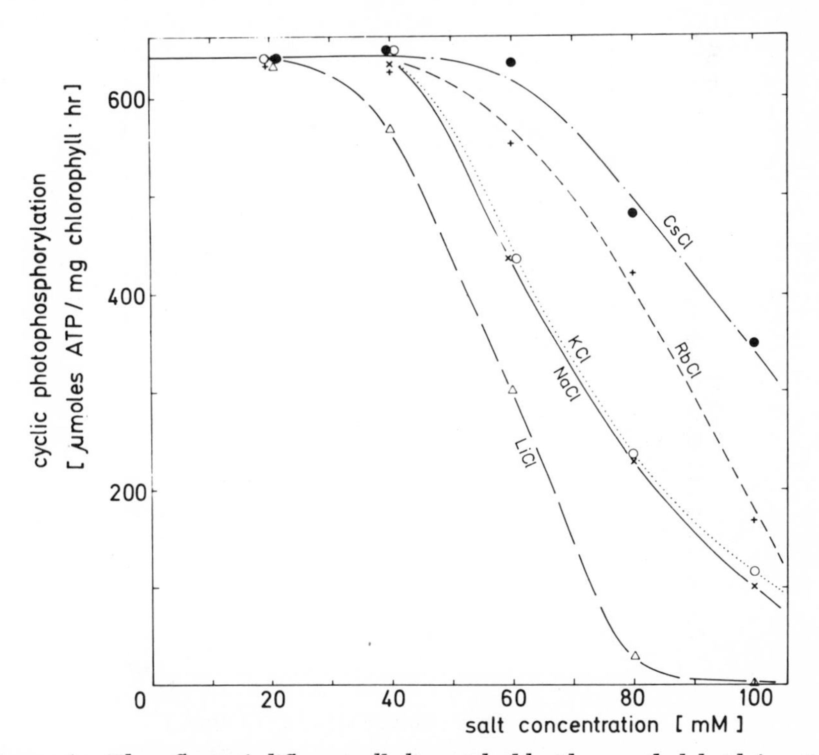

Figure 4. The effect of different alkali metal chlorides on thylakoid function during freezing to −20°C. Washed thylakoids were suspended before freezing in a solution containing 0.1 M *sucrose and alkali metal chlorides, as indicated. The sucrose served as cryoprotectant and was added to prevent freeze-inactivation of the membranes in the presence of low salt concentrations. The suspensions were slowly frozen for 3 hours at −20°C. After thawing in a water bath at room temperature, the activity of cyclic photophosphorylation was measured. For experimental conditions, see the legend for Fig. 2.*

in that order: $Li^+ > Na^+ > K^+ > Rb^+ > Cs^+$. At 0°C, high concentrations of all alkali metal chlorides produced comparable membrane inactivation (*24*). Similar results have been obtained during freezing of red blood cells (*25*). Chlorides of divalent cations such as $CaCl_2$ produce more drastic membrane inactivation than chlorides of alkali metals (Figure 5). $MgCl_2$ had an intermediate effect. However, low concentrations of Mg^{++}, rather than being toxic, are actually necessary for photophosphorylation of thylakoids.

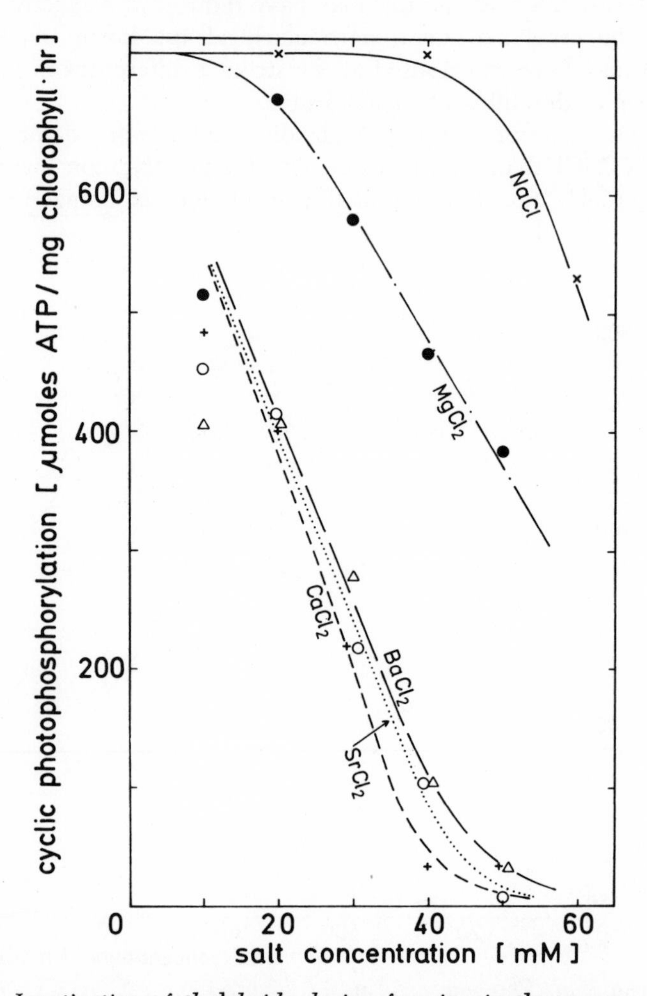

Figure 5. Inactivation of thylakoids during freezing in the presence of chlorides of different divalent and monovalent metals. Washed thylakoids were suspended before freezing to −20°C in a solution containing 0.1 M *sucrose as cryoprotectant and various chlorides at concentrations indicated on the abscissa. After thawing, the activity of cyclic photophosphorylation was measured. For experimental conditions, see legend for Fig. 2.*

ANIONS. By holding the cation constant and varying the anion, the effects of different anions on thylakoids during freezing or at 0°C can be compared. As in the case of cations, low concentrations of different anions are not destructive if occasional specific effects on photophosphorylation, such as that exerted by sulfate (*26*), are disregarded (Figure 6). The membrane toxicities of different anions vary widely at high concentrations. Among the halogenides, fluoride is tolerated even at rather high concentrations. Toxicity increases in the order of chloride, nitrate, bromide, and iodide [Figure 6 and (*24*)]. With red blood cells, the order of anion cryotoxicity also increased from chloride to iodide (*25*). However, it is important to emphasize that membrane inactivation is slight when the salt concentration is low, even for those salts that are highly detrimental at high concentrations. For example, chloride at a low concentration is actually necessary for thylakoid function (*27*).

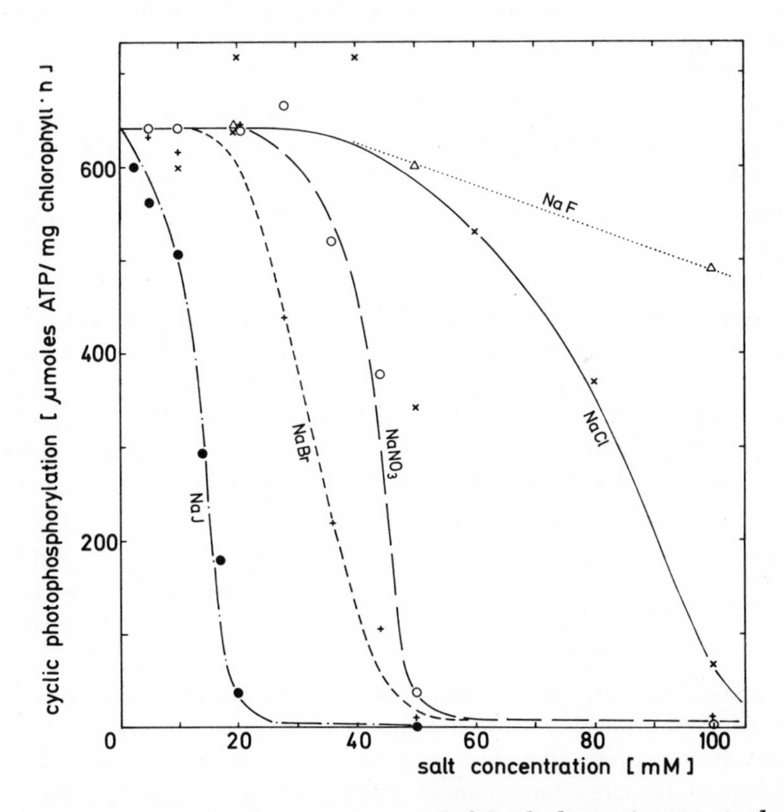

Figure 6. Cryotoxic effects of anions on thylakoids during freezing in the presence of different sodium salts. Washed thylakoids were suspended in a solution containing 0.1 M *sucrose and various sodium salts at concentrations indicated on the abscissa. For experimental conditions see legend for Fig. 2.*

Organic Salts. Organic salts, as is true of inorganic salts, are usually nontoxic at very low concentrations. At higher concentrations, their membrane toxicity varies greatly. Freezing red blood cells in the presence of sodium acetate produces hemolysis (*25*). Thylakoids, however, are protected against freeze-inactivation by sodium acetate (*21*). This effect of sodium acetate is markedly different from the detrimental effect that sodium chloride usually has on thylakoids during freezing. Obviously, sodium acetate is a cryoprotectant for thylakoids, but not for red blood cells. Sodium citrate, pyruvate, malate, and tartrate are other examples of organic salts which can, at least to some extent, prevent the freeze-inactivation of thylakoids (*21*). In these instances, low concentrations are less protective than high concentrations.

In contrast to these salts, sodium succinate increases thylakoid damage during freezing if present as the only solute. As will be discussed later, the situation becomes more complicated if significant concentrations of other solutes are also present with sodium succinate during freezing. Membrane inactivation by high concentrations of sodium succinate occurs even at 0°C (*21*).

Salts of weak organic acids that are soluble in lipids are also injurious to thylakoids. Examples are the salts of phenylpyruvic acid (Figure 3) and caprylic acid (*28*). These salts, even if present at very low concentrations, cause extensive membrane inactivation during freezing, if cryoprotectants are absent. At 0°C, moderate concentrations of these salts will slowly inactivate thylakoids.

Amino Acids. As is true of organic acids, amino acids can either prevent inactivation of thylakoids by freezing or they can aggravate the situation. Some of them, for instance glycine, serine, glutamate, or aspartate, promote injury if present as the only major solutes during freezing. However, the same amino acids can be protective if certain other solutes are also present (*28*). The reason for this behavior, which is also observed with succinate, will be considered later.

Proline, threonine, or γ-aminobutyric acid can protect thylakoids against inactivation during freezing. Amino acids with apolar side chains such as phenylalanine, leucine, or valine always contribute to thylakoid inactivation during freezing.

Proteins. Usually, thyalkoids at 0°C can tolerate soluble proteins at high concentrations but these proteins are usually not protective during freezing. Remarkable exceptions are some of the proteins found in different organs of frost-resistant plants. These proteins contain a high percentage of hydrophilic amino acids, are heat stable, have molecular weights ranging between 10,000 and 20,000 daltons, and are believed to play an important role in frost hardiness (*29, 30, 31*).

Neutral Solutes. Among the neutral compounds, sugars occupy a dominant position as far as physiological importance is concerned. Thylakoids tolerate very high concentrations of sugars at 0°C (*32*). Freezing in the presence of sufficiently high concentrations of highly soluble sugars such as raffinose, sucrose, glucose, or ribose does not lead to membrane damage (*15, 33*). Sugars thus possess cryoprotective properties. The same is true for sugar alcohols. However, protection of thylakoids against freezing damage can be observed only when eutectic freezing is avoided (*34*). Mannitol, for instance, is not effective in preventing the inactivation of thylakoids during freezing to low temperatures because it crystallizes easily (*7*).

Glycerol is used as a cryoprotective agent in the freezing of spermatozoa and red blood cells (*35, 36, 37*). Another compound that has found practical application in the freeze-preservation of cells is dimethylsulfoxide. Both glycerol and dimethylsulfoxide can prevent inactivation of thylakoids during freezing. Interestingly, ethylene glycol and even methanol or ethanol have cryoprotective properties (*38*), but only if they are not the predominant solutes (*13*).

During hardening, many plant cells produce their own internal supply of cryoprotective compounds. In those instances, nonpenetrating soluble sugars such as raffinose or sucrose are often accumulated, and resistance to freezing increases.

Membrane Damage During Freezing

Mechanical Damage Piercing of biomembranes by growing ice or other crystals is a self-evident means of damage. During fast freezing, intracellular ice formation appears to cause mechanical damage. Also, eutectic freezing leads, even in the presence of a high concentration of a cryoprotective solute, to compete inactivation, which is presumably due to mechanical damage (*7*). During slow, natural freezing, mechanical membrane injury is the exception, not the rule.

Solute Injury. As has been outlined above, a number of natural cell constituents are potentially harmful. In the absence of freezing and at physiological concentrations, they are innocuous and may even be essential for the normal functioning of cellular metabolism. However, during freezing their concentrations can rise to damaging levels causing irreversible alterations in membranes.

MEMBRANE AGGREGATION. Isolated thylakoids form stable suspensions at neutral pH, since the membranes carry a net negative charge. During thawing of thylakoid suspensions which have been frozen in salt solutions the membranes aggregate and precipitate. This indicates a reduction in the net charge of the membranes. Precipitation does not occur when the membranes are frozen in sufficiently concentrated solutions of sucrose or other cryoprotective solutes. Usually, precipitation indicates membrane damage.

PERMEABILITY CHANGES. As has been mentioned, exposure of red blood cells sufficiently concentrated NaCl solutions at 0°C or during freezing will result first in shrinkage then hemolysis. Obviously, the cellular membrane becomes leaky and permits passage of solutes. In thylakoids, ATP synthesis during photophosphorylation is inactivated by freezing. It is widely accepted that energy conservation in mitochondria and chloroplasts requires membranes with a low proton permeability. Proton transport across vesicular membranes, which is coupled to electron flow (*39*), leads to the formation of transmembrane proton gradients, the energy from which is believed to be used for the endergonic synthesis of ATP (*40*). Indeed, in thylakoids which had been inactivated by freezing, light-induced transmembrane proton gradients are decreased or abolished, even if electron transport is little affected (*41, 42*). When the membranes have suffered only mild damage by freezing, electron transport may actually be enhanced, while photophosphorylation is lost (*5, 22, 28*). Uncoupling of phosphorylation from electron transport is thus observed.

It is not only proton permeability that changes after freezing. Functional thylakoids respond osmotically to the addition of many solutes (*41, 43*). They shrink in hypertonic solutions of sucrose or NaCl and expand in hypotonic solutions. Their permeability to cations (*44*) and hydrophilic neutral molecules of moderate or large size (*43*) is low. They exhibit somewhat greater permeability to some anions such as chloride (*45, 46, 47*). Glycerol (*43*) or ethylene glycol penetrate rapidly. After freezing damage, membrane permeability increases indiscriminately. Osmotic responses are no longer observed and the membranes appear collapsed when viewed with an electron microscope (*41*).

ELECTRON TRANSPORT. It has been mentioned that mild freezing can inactivate photophosphorylation and at the same time stimulate light-dependent electron transport in thylakoids (*5*). Electron flow from water through both photosystems is increased in damaged thylakoids. However, after freezing in the presence of comparatively high levels of substances that are potentially toxic to membranes, electron transport from water decreases and the water-splitting system becomes inactivated (*28*). Still, electron transport from a donor such as ascorbate through photosystem 'I' may be much greater in such membranes than it is in functional thylakoids. Only after very severe freezing stress is electron transport through photosystem 'I' decreased (*48*). These observations show that the extent of membrane damage during freezing depends on the freezing conditions and the solute environment.

PROTEIN RELEASE. Biomembranes consist of lipids and proteins. The latter may be subdivided into so-called intrinsic and extrinsic proteins (*49*). Intrinsic proteins supposedly are integrated into the membrane phase primarily by the hydrophobic interaction with lipids. Extrinsic proteins are attached to the membranes. Ionic interactions are believed to be important in the binding of extrinsic proteins. When these proteins dissociate from the membrane, they may be sufficiently hydrophilic to be soluble in the aqueous phase. When freeze-aggregated thylakoids are sedimented, a number of membrane proteins are found in the supernatant fluid. Among them are catalytic proteins involved in energy conservation and electron transport (*42, 48*). The total amount of proteins released depends on freezing conditions and the solute environment, but may be as much as 5% of the total membrane protein (*48*). When frozen in the presence of a cryoprotective solute, at a sufficient concentration, thylakoids remain functional and do not release proteins in significant amounts. Protein release thus accompanies membrane injury and, in fact, is an indication of such injury.

Freezing is not the only cause of protein release. Proteins dissociate from the membranes during inactivation of thylakoids by exposure to high concentrations of salts at 0°C (*48, 50*). The pattern of membrane proteins released by freezing is similar to the pattern of proteins found in the supernatant fluid from membranes which have been exposed to a high salt concentration at 0°C (Figure 7). Usually, eight bands are clearly apparent in electropherograms of proteins released from thylakoids during freezing in the presence of NaCl. Seven further bands are faint and they constitute minor components. The molecular weights of the released polypeptides range between 15,000 and 60,000 daltons.

Differences in the general pattern of released proteins occurred when thylakoids were frozen in solutions containing different cryotoxic compounds. More proteins were released when thylakoids were frozen in the presence of NaBr or KBr rather than NaCl. This probably can be attributed to the greater cryotoxicity of the bromide anion. The pattern of polypeptides produced when thylakoids were frozen in the presence of sodium phenylpyruvate, sodium caprylate, or isoleucine differed greatly in a qualitative manner, and also to some degree in a quantitative manner, as compared to the pattern obtained when the membranes were frozen in a NaCl solution. The organic cryotoxic solutes containing apolar side chains released much more of polypeptides 5 and 6 than did the inorganic salts. The interaction of sodium phenylpyruvate with the membranes also can be directly observed at 0°C or 20°C by monitoring slow changes in light scattering of the membranes brought about by the salt (*51*).

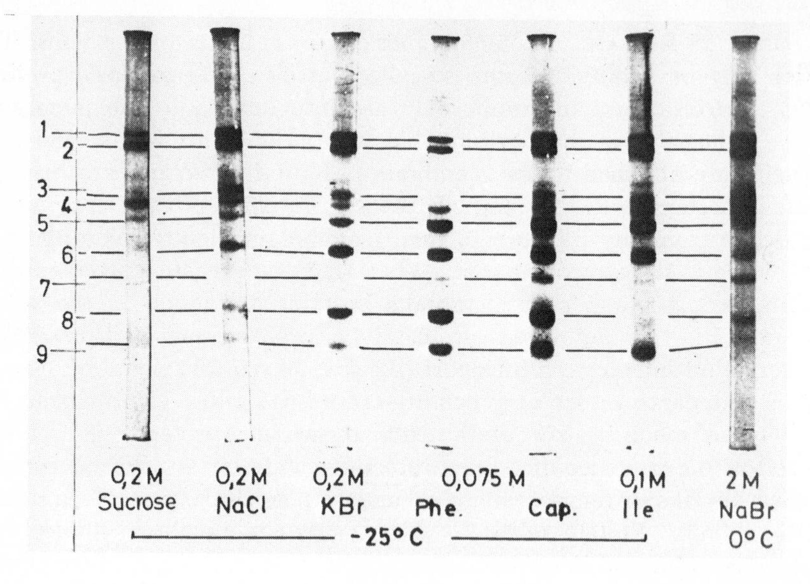

Figure 7. Electrophoretic patterns of proteins which are released from thylakoids during freezing or exposure to 0°C in the presence of various solutes. The solute concentration before freezing is indicated under the gels. Freezing time was 3 hours at −25°C. After thawing, supernatant fluids from membranes were treated with sodium dodecylsulfate (SD) and mercaptoethanol then subjected to gel electrophoresis. Phe is sodium phenylpyruvate, Cap is sodium caprylate, Ile is isoleucine. From Volger, Heber, and Berzborn (48).

In view of the partially nonpolar properties of the phenylpyruvate and caprylate anions and of the lipid solubility of their protonation products which are in equilibrium with the anions, it is likely that these organic salts not only release extrinsic hydrophilic proteins which are attached to the membranes, but also affect hydrophobic bonding within the membrane structure. Such effects cannot be seen in protein release experiments because apolar proteins remain insoluble.

Some of the proteins released during freezing have been identified either by immunoelectrophoretic analysis or by co-electrophoresis of pure proteins or polypeptide chains. Bands 1, 2, 4, 8, and 9 contain the α, β, γ, δ and ϵ-subunits of the coupling factor CF_1, respectively. The coupling factor is responsible for the synthesis of ATP in photophosphorylation. Still, the loss of photophosphorylation during freezing involves more than just the loss of the coupling factor (*15, 41, 42, 52*). Since the ion gradients thought to drive the endergonic synthesis of ATP can be maintained only by membranes having a low ion permeability, the observed loss of semipermeability during freezing is by itself a sufficient cause for the inactivation of photophosphorylation.

From Figure 7 it is apparent that the subunits of CF_1 appear in solution at very different ratios depending on the nature of the cryotoxic solute present during freezing. An antiserum to the native coupling factor did not react with the supernatant fluids derived from membranes damaged by freezing. If CF_1 had left the membrane as an intact molecule and had subsequently dissociated into subunits as might be expected from its cold lability in solution (*53*), it would have given rise to band patterns of a uniform intensity distribution. The observed intensity distribution, which was very different in the presence of different cryotoxic solutes, suggests instead that the molecule had disintegrated on the membrane and had released only some of its subunits. Indeed, the δ-subunit of CF_1 was shown by specific antisera to be present in particularly large amounts in supernatant fluids from membranes frozen in the presence of sodium caprylate. When the membranes were frozen in the presence of isoleucine, very little of the δ-subunit, but a large proportion of the α and β-subunits, was released into solution. It should be noted that the known cold-lability of the coupling factor is normally exhibited only in solution, and not when the molecule is integrated into the membrane. The observed disintegration of the molecule during freezing therefore appears to be predominantly a solute effect, though temperature may also play a role (*15*).

Band 2 of Figure 7 was occasionally seen to contain a second component in addition to the β-subunit of CF_1. This component is probably the large subunit of carboxydismutase which sometimes tends to attach to thylakoids although it is a soluble enzyme. The small subunit of carboxydismutase, and plastocyanin, which was identified by a specific antiserum, are located in the area of band 9. The protein of band 3 is probably ferredoxin-NADP-reductase. The bands 5 to 7 remain unidentified, as are the 7 to 9 faint bands which are not visible in Figure 7.

While considerable protein was released when the membranes were frozen and inactivated in the presence of salts or isoleucine, some protein loss is also apparent in the sucrose experiment of Figure 7. During freezing in sucrose solution, the membranes remained functional. As in the salt experiments, the α- and β-subunits of the coupling factor were prominent among the components released in the sucrose experiment. However, there is good reason to assume that the partial loss of the coupling factor, which sometimes occurred in the presence of the cryoprotective agent sucrose, was attributable to different causes than those which led to destruction of the coupling factor during freezing in the presence of cryotoxic salts. The coupling factor can be solubilized and removed from the thylakoids by treatment with EDTA, which effectively complexes divalent cations (*54*). Even washing the membranes with salt-free sucrose solutions can detach the coupling factor (*55*). Indeed, membranes washed less carefully than the membranes of the sucrose experiment of Figure 7 showed very little loss of protein during freezing in the presence of sucrose.

The possibility should be considered that protein loss during inactivation of thylakoids by freezing results from an increase in membrane permeability, which would permit the leakage of intrathylakoid proteins through the membranes into the medium. However, among the identified proteins released from thylakoids, only plastocyanin is located on the inside of the membranes. Coupling factor CF_1 and NADP reductase are attached to the outside of the thylakoids (*56, 57, 58, 59, 60*). When supernatant fluids derived from membranes that had been inactivated by freezing were used as immunogens, the resulting antisera agglutinated intact thylakoids. Since antibodies are large hydrophobic molecules which cannot penetrate biomembranes, this also shows that antigens are located on the matrix side of thylakoids (*48*).

PROTEIN RELEASE IN RELATION TO LOSS OF MEMBRANE FUNCTION. When thylakoids are frozen in the presence of sucrose, membrane function is preserved. If a cryotoxic salt such as NaCl is also present, retention of membrane functionality during freezing depends on the ratio of sucrose to salt (*5*). Loss of cyclic photophosphorylation is the most sensitive parameter of membrane inactivation. Photophosphorylation is largely lost before significant protein release from the membranes can be detected (Figure 8). Since photophosphorylation requires membranes with un-

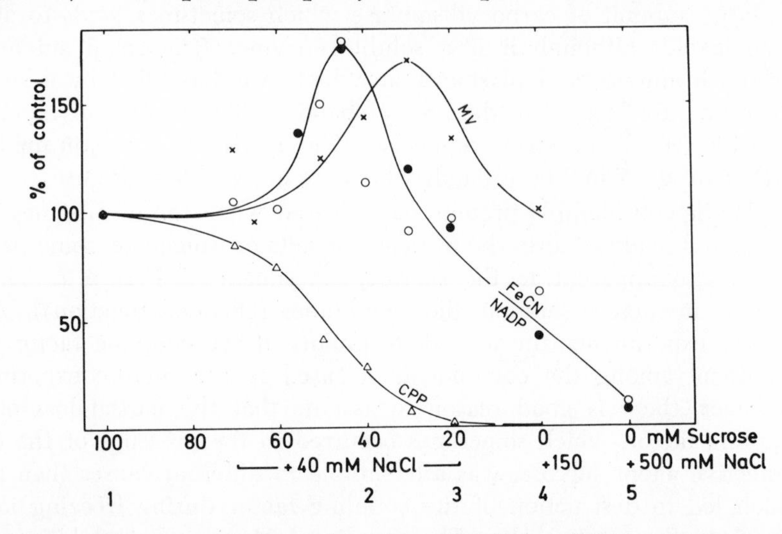

Figure 8a. Photosystem-I-dependent phosphorylation (CPP), photosystem-I-dependent electron transport (MV, methylviologen reduction in the presence of an electron donor system), and electron transport through photosystems II and I (NADP and ferricyanide reduction) in thylakoids after freezing for 3 hours to −25°C in solutions containing different ratios of sucrose to NaCl

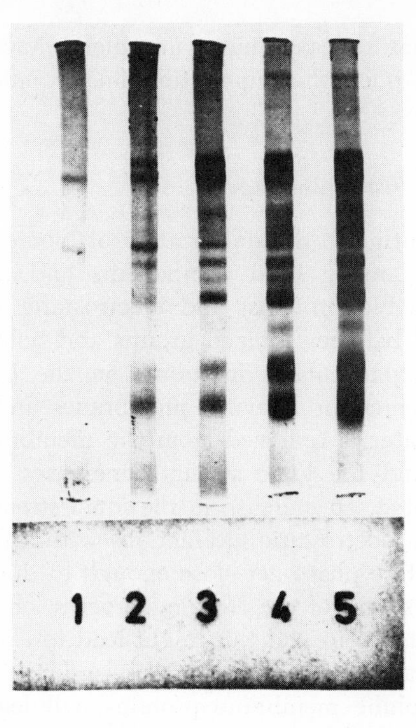

Biochimica et Biophysica Acta

Figure 8b. Polypeptide patterns of proteins, which were released from thylakoids during freezing. The numbers 1 to 5 relate to the conditions shown along the abscissa of Fig. 8 (A). From Volger, Heber and Berzborn (48).

changed permeability characteristics, this suggests that permeability changes occur before much protein dissociates from the membrane. Thus protein release indicates an advanced rather than an initial state of membrane damage. As photophosphorylation decreases, the rate of electron transport increases. This shows that the loss of components of the electron transport chain from the membranes is not yet critical, even though protein release is already significant. As the ratio of salt to sucrose increases, membrane damage becomes more severe and electron transport is decreased. At the same time, the dissociation of proteins from the membranes increases considerably. As has been mentioned, components of the electron transport chain such as plastocyanin and NADP reductase are among the released proteins (*42, 48*). Indeed, Steponkus et al. (*15*) have shown that loss of electron transport can be decreased by exposing thylakoids to a high concentration of plastocyanin during freezing. This minimizes loss of plastocyanin from the membranes. However, the conclusion is premature that protein release is more closely related to inactivation of electron transport than to the freeze-induced changes in membrane permeability which cause loss of normal membrane function.

It is rather likely that a weakening of intramembrane interactions which first causes loss of photophosphorylation finally culminates in protein dissociation.

Mechanism of Membrane Damage

Factors contributing to the stabilization of biomembranes are hydrophobic interactions among lipid components and between membrane lipids and hydrophobic proteins, and electrostatic interactions among membrane ions and between ionized groups and polar molecules. Electrostatic forces are particularly important in the binding of extrinsic proteins and in interactions between membranes and the solute phase. During freezing, water is removed from the membrane suspension and converted to ice, and the ionic strength increases. Since electrostatic forces are not oriented, an increase in the ionic strength of the medium will finally suppress electrostatic interactions within the membrane, provided ions of the solute phase get close enough to the ionized membrane sites. In the initial stages of the freezing process, changes in membrane structure are likely to occur and this might lead to changes in membrane permeability. As damage progresses and a sufficient number of bonds are cleaved, hydrophilic membrane proteins will leave the membrane. This process is shown below for a situation involving interaction of NaCl with noncovalent bonds linking two protein molecules in a single membrane:

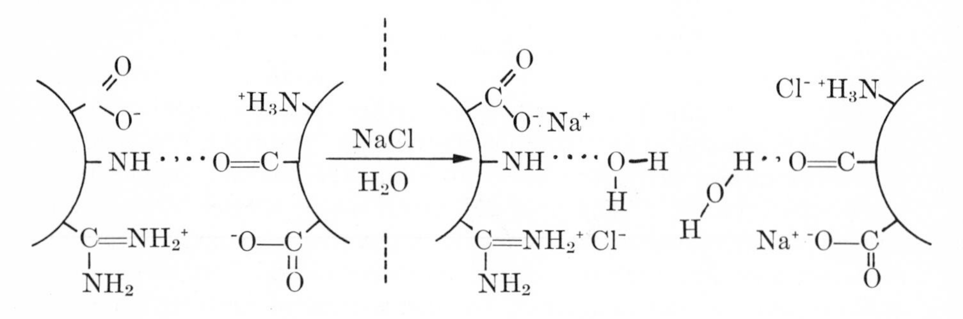

Aggregation of membranes will occur if binding of counterions reduces the net charge of membranes, thereby facilitating noncovalent interactions between different membranes.

It is appropriate to consider the toxicities of different ions toward membranes. During freezing of thylakoids, anion toxicity decreases in the order $I^- > Br^- > NO_3^- > Cl^- > F^- >$ acetate . This is reminiscent of the Hofmeister lyotropic power series, which was originally observed with regard to denaturation of euglobulins, then for blood clotting, then for the hydrothermal shrinkage temperature of collagen and other effects (*61, 62, 63*). The similarity between the anion series that functions during the freeze-inactivation of thylakoids and the Hofmeister series lends support to the idea that salt inactivation of biomembranes occurs through interference with polar binding of membrane components. Larsen and Magid (*64*) measured heats of transfer of a variety of salts from water to solutions of micelle-forming surfactants. The micelles may, for our purpose, serve as simple models of membranes. Binding of anions to cationic micelles, which involves competition with and replacement of original counterions, was strongest for anions having a small Stokes' law hydrated radius and decreased with increasing radius. The order of anion binding as derived from ΔH trans was $Br^- > NO_3^- > Cl^- > F^- >$ acetate$^-$ $>$ citrate^{3-}. This Hofmeister power series is virtually identical with the order of membrane toxicities exerted by anions in freezing experiments with thylakoids. It is therefore concluded that cryotoxic anions exert membrane effects by competing with membrane anions for cationic binding sites on the membrane, thereby suppressing intramembrane interactions.

Membrane inactivation depends on how closely anions can approach cationic binding sites. Poorly solvated ions show the strongest binding. They are also known to be the most effective protein denaturants (*65*). The Stokes' law hydrated radius of the toxic bromide anion is about 1.2 Å, that of the relatively nontoxic fluoride about 1.6 Å and that of the cryoprotective acetate anion 2.2 Å. Biological membranes usually appear in thin sections as three-layered structures 60 to 100 Å thick. In view of this relatively large cross section, accessibility of binding sites becomes of obvious importance.

Interestingly, the envelope of intact chloroplasts has an anion permeability, which follows the order of anion toxicity to thylakoid membranes during freezing. The iodide anion penetrates rapidly and is followed in order by bromide, chloride, and fluoride (Figure 9). Acetate anions penetrate chloroplasts slowly (*46, 66*). Since the rate of membrane penetration by, for instance, chloride is proportional to its concentration gradient, and since no competition apparently exists among various penetrating anions, the anions diffuse across the membranes rather than being transferred by a carrier mechanism. The anion permeability of thylakoids seems to be lower than that of the chloroplast envelope, but here again iodide was found to be the anion that penetrates most rapidly (*67*). Membrane penetration by the anions shows that not only external but also internal membrane sites are accessible for interaction.

For cations, the observed order of toxicity during freezing of thylakoids is Sr^{++}, Ca^{++}, $Ba^{++} > Mg^{++}$ and $Li^+ > Na^+$, $K^+ > Rb^+ > Cs^+$. At 0°C, high concentrations of monovalent alkali metal cations exhibited

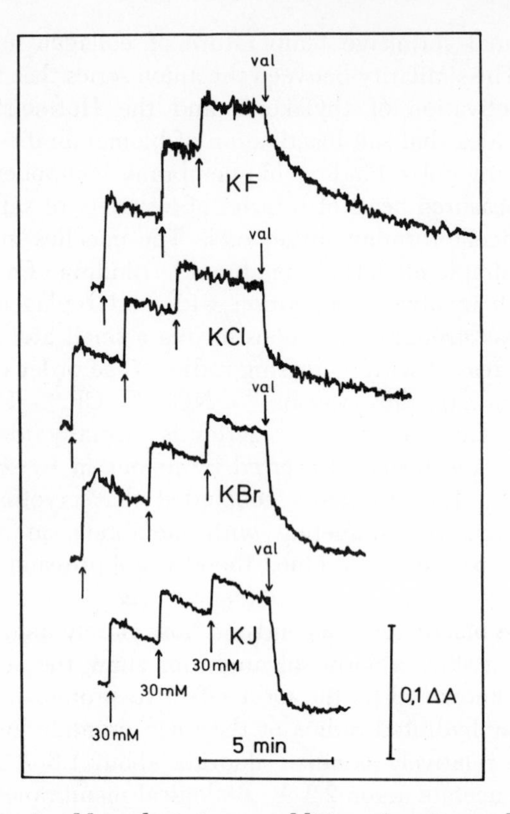

Figure 9. Change in chloroplast size on addition of various substances. Halogenides were added in 30 mM increments, which resulted in shrinkage. These additions were followed by valinomycin (val) at a concentration of 2 μM, which resulted in expansion. Changes in chloroplast size at 20°C were monitored by changes in the apparent absorbance of the chloroplast suspension at 535 nm (43). Note different slopes of the absorbance decrease seen on addition of valinomycin, which increases the K^+ permeability of the chloroplast envelope. As in the presence of the antibiotic, K^+ diffusion is not limiting the rate of salt uptake, different slopes indicate different anion fluxes. For experimental conditions see *Ref.* 91.

only small differences in their membrane toxicities (*24*). The order of membrane toxicities of cations as observed during freezing of thylakoids is similar to the order of cation effects in the lyotropic power series of Hofmeister and others (*61, 62, 63*). It is also similar to the order in which cations denature proteins, that is, strongly hydrated species such as Li^+ are better denaturants than weakly hydrated species such as Cs^+ (*65*). The correlation between cationic toxicity to membranes and cationic binding to anionic surfactants is not as impressive as in the case of anions (*64*). Kuntz and Taylor (*65*) have already noted that arguments similar to those used to explain differential anion binding are not likely to apply to cationic effects on denaturation. Still, in agreement with the order of ion toxicity to membranes, the enthalpy for the transfer of salt to a solution of sodium dodecylsulfate was more negative for $CaCl_2$ than for $MgBr_2$, and that for $MgBr_2$ was more negative than that for alkali metal bromides (*64*).

It has been mentioned that thylakoids are more sensitive to salts of certain organic acids such as phenylpyruvic acid and caprylic acid than to salts of inorganic acids (*28*). Differences in the release of proteins (*48*), and different poylpeptide patterns of released proteins (Figure 7), suggest differences in the action of inorganic and organic anions. The latter bind to membranes electrostatically as well as by other means. This has been verified by calorimetric measurements of the binding energy of the tosylate anion to cationic surfactants (*64*). Depending on their p*K* values, some low proportion of the phenylpyruvate and caprylate anions are protonated even at neutral pH. The protonated species are lipid soluble and on freezing their concentrations increase together with the concentrations of ions. The lipid-soluble material will accumulate in the lipid part of the membrane phase and disturbs hydrophobic interactions in the membrane. As a consequence, formation of protuberances and disintegration of membranes can be easily observed with a microscope when high concentrations of sodium caprylate are added to thylakoids. Hydrophobic damage to membranes will not, however, be very apparent in protein release experiments, since apolar proteins and lipids cannot be expected to leave the membrane phase even if the membrane structure is seriously disturbed by freezing.

Phenylpyruvate and caprylate are nonphysiological salts. Thus, the physiological relevance of observations made with these compounds might be questioned. However, plant cells contain many solutes which have effects on thylakoids similar to those exerted by phenylpyruvate or caprylate. Included among these compounds are the amino acids phenylalanine, valine, leucine, and isoleucine and a number of phenolic substances.

Membrane Protection

Colligative Protection. The principles of colligative protection were first outlined by Lovelock (*35*) for the red blood cell. These principles are also valid for the thylakoid system (*14, 21, 68*). If only one solute is present in a membrane suspension, its concentration, regardless of its initial concentration, will rise during freezing to a level determined solely by the final freezing temperature. If the solute is a cryotoxic compound, this final level may be sufficient to cause membrane inactivation. When several solutes are present and only one is a cryotoxic solute, the same total osmolar solute concentration will be obtained when solid–liquid equilibrium is achieved at the same subfreezing temperature used in the first case. However, in the second case several solutes contribute to the final solute concentration, and the cryotoxic solute constitutes only a fraction of the total osmolar concentration. Its fractional concentration depends, simply, on the ratios of the various solutes existing in the original sample. If the fraction of the cryotoxic solute is small, its concentration may not reach a damaging level during freezing. Even if the nontoxic solutes that "dilute" the cryotoxic compound do not exert any direct influence on the membranes, they will nonetheless act as cryoprotective agents.

These relations make it possible to explain the paradoxical observation that cryotoxic solutes such as NaCl can sometimes act as membrane protectants during freezing (*21*). Thylakoids suspended in a medium containing, for instance, sodium succinate as the predominant solute are inactivated by freezing, because high concentrations of succinate are not tolerated by the membranes (Figure 10, left part, II a,b,c). If, however, increasing concentrations of NaCl are added to the membrane–succinate system, a range of NaCl concentrations is encountered where freezing does not result in membrane damage. At these particular ratios of succinate to chloride, neither of the anions will reach damaging levels during freezing and protection is observed. The solute ratios, not the initial concentrations, determine when protection occurs. If the ratio of NaCl to succinate is further increased, freezing will raise the concentration of NaCl to a level that is damaging.

Such observations show that the term "cryoprotective agent" has a very loose meaning and does not necessarily imply that a compound plays any active role in membrane stabilization. When a soluble sugar (Group I, Figure 10) is present in the thylakoid suspension, freezing will not cause membrane damage, as long as the ratio of sugar to NaCl does not fall below a critical threshold value, because even high concentrations of sugars are tolerated by the membranes.

Specific Protection. LOW MOLECULAR WEIGHT SOLUTES. Colligative protection is nonspecific. Any solute which does not damage a biomembrane must, just by its presence, reduce the concentration of a potentially cryotoxic solute during freezing as compared to the concentration the cryotoxic solute would attain in the absence of a second solute. On an osmolar basis, different neutral solutes should be equally effective. However, the facts differ from this expectation. On an equi-osmolar basis, raffinose is a better cryoprotectant than sucrose is superior to glucose, and glucose is more effective than glycerol (*15, 33, 34*). Similar deviations from "ideal" colligative behavior have been observed for other solutes (*68*). To explain the differences, it is necessary to either introduce

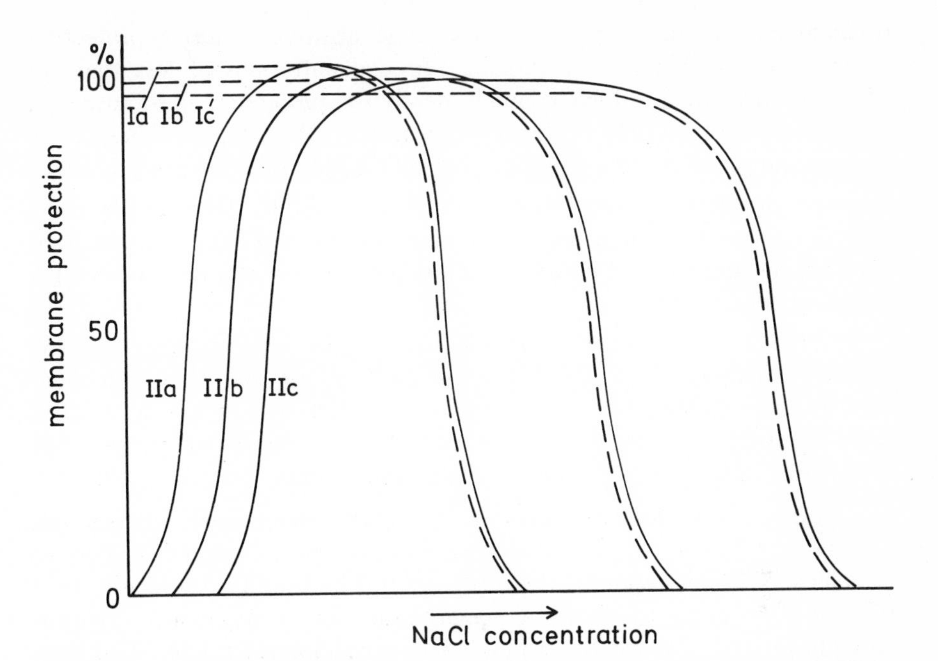

Figure 10. Thylakoid protection by two different groups of compounds as a function of the concentration of a potentially cryotoxic solute **such as NaCl.** *Examples of solutes belonging to group I are soluble sugars, sugar alcohols, proline, threonine. Representatives of group II include sodium succinate, glutamate and asparate. a, b, c, are different concentrations increasing from a to c. The osmolar ratio of a given compound from group I or II versus NaCl, that is needed to produce 50% inactivation of the membrane during freezing, differs depending on the compound (not indicated in this figure).*

osmotic coefficients or to assume that solutes have direct solute-specific effects on membranes in addition to their colligative effects. Direct stabilization can indeed be observed in the absence of freezing, when colligative dilution of membrane-toxic solutes is not possible. For example, thylakoids suspended in NaCl solutions at 0°C are inactivated faster in the absence of sucrose than in its presence (*21*). Furthermore, dimethyl sulfoxide can protect unfrozen ovary cells of the Chinese hamster against damage by a hyperosmotic salt solution (*69*). Molecular details of specific membrane protection are not yet known. It is possible that effects on water structure are involved.

Cryoprotective Proteins. In general, soluble proteins are either weakly effective or ineffective for preventing the freeze-inactivation of thylakoids suspended in dilute salt solutions. This is not unexpected since low concentrations of high molecular weight compounds such as proteins cannot significantly reduce the freeze-concentration of cryotoxic solutes by colligative action. However, some proteins extracted from frost-resistant plants can, even at very low concentrations, exert a protective effect on thylakoids during freezing. At concentrations of less than 40 μM, these proteins produce as much membrane protection as sucrose at a concentration of 30 mM (calculated from refs. 31 and 70). These facts alone make protection on a colligative basis highly unlikely for proteins.

The proteins are heat stable and water soluble. The amino compositions of two of them have been determined, and they contain high percentages of polar amino acids and low percentages of amino acids with nonpolar side chains.

Little is known concerning mechanisms by which these proteins prevent inactivation of thylakoids during freezing, but they somehow contribute to membrane stabilization. They act with some specificity, since cryoprotective proteins from spinach not only fail to protect red blood cells during freezing but are actually injurious.

Ability of Cryoprotectants to Penetrate Membranes. Preserving normal membrane properties during freezing poses special problems. Lovelock (*71*) has reported that only neutral hydrophilic molecules of a size small enough to penetrate membranes were successful cryoprotectants for red blood cells suspended in physiological saline. This suggests that not only the outer but also the inner side of the cellular membrane requires protection. More recent work (*92*) has shown that the inner side is less sensitive to freezing damage than the outer one, perhaps because solutes of the inner phase contribute to protection. Provided the interior of a cell contains an excess of a cryotoxic solute, protection is possible only if it can leak out during freezing to be colligatively diluted by a nonpenetrating cryoprotectant located outside, or if a penetrating cryoprotectant can enter the cell. If both cannot penetrate, they must be present on the same membrane side for protection to become possible.

In plant cells, accumulation of soluble sugars has often been noticed during hardening and this has been suggested as a factor in frost hardiness (*12, 14, 68*). Extracellular addition of sugars to frost-sensitive cells fails to protect against freezing injury (*72, 73*) except in cases where sugar uptake was substantial (*74, 75*). In view of this, it is surprising that thylakoids suspended in solutions containing cryotoxic solutes can be protected against freeze-inactivation not only by penetrating solutes such as glycerol or dimethylsulfoxide, but also by relatively large sugars such as glucose, sucrose, or raffinose (*15, 33*), which are normally regarded as nonpenetrating (*43*). As mentioned above, even some proteins with molecular weights of 10,000–20,000 daltons protect thylakoids against freezing damage, although complete protection is not observed unless other cryoprotectants are present(*31*).

Protection of thylakoids by large molecules might be explained in several ways. Perhaps the thylakoid membrane is sensitive to solute injury only on the outer side, or the intrathylakoid space contains, even after long incubation, little of the added cryotoxic solute, or penetration of the membrane by normally nonpenetrating solutes becomes possible under freezing conditions (*76*).

The fact that cryoprotective proteins alone cannot provide complete protection to thylakoids during freezing, even when present at saturating concentrations, can be regarded as evidence that not only the outside but also the inside of the thylakoid membrane requires protection. When frozen in a sucrose solution of sufficient concentration, thylakoids are completely protected. Although sugars, when first added to thylakoid membranes, produce an osmotic response as expected from van't Hoff's law, they apparently leak into the intrathylakoid space during freezing (*76*), thereby providing protection to the inner side of the membrane. It is not known whether cryotoxic solutes leak concurrently out from the intrathylakoid space. Obviously, the permeability properties of biomembranes have an important role in membrane protection. Since the permeability properties of a membrane depend on membrane structure, any structural changes should influence membrane survival during freezing.

Protection by Changes in Membrane Structure. Development of frost hardiness in plants is often accompanied by an increase in membrane lipids, particularly phospholipids (*77, 78, 79*). An increase in the degree of unsaturation of the fatty acid components of phospholipids also has been observed by some (*78, 79*) but not by all investigators (*77*). Attempts to correlate increases in membrane lipids with hardiness of thylakoids from cabbage leaves have so far failed (*80*). Also, what effect an increase in membrane lipids may have on hardiness is not yet understood.

It is known that hardy plant cells exhibit increased permeability to water (*12, 81, 82, 83*). In some cases, cytological and ultrastructural differences in chloroplasts of hardy and nonhardy leaves have been reported (*15, 84, 85*). These observations require interpretation.

By means of suitable membrane-active additives, we have attempted to modify the structure of isolated thylakoids and thereby increase membrane permeability. Changes in resistance of the membranes to freezing were then determined. Figure 11 shows the effect of sodium caprylate on the permeability of thylakoids to protons. In the absence of the compound, illumination causes proton transfer across the thylakoids and acidification of the intrathylakoid space. A fluorescent weak amine was used to monitor formation of the proton gradient (*86*). When the light was turned off, protons moved slowly back across the thylakoid membrane. Transport followed first-order kinetics. In the presence of caprylate

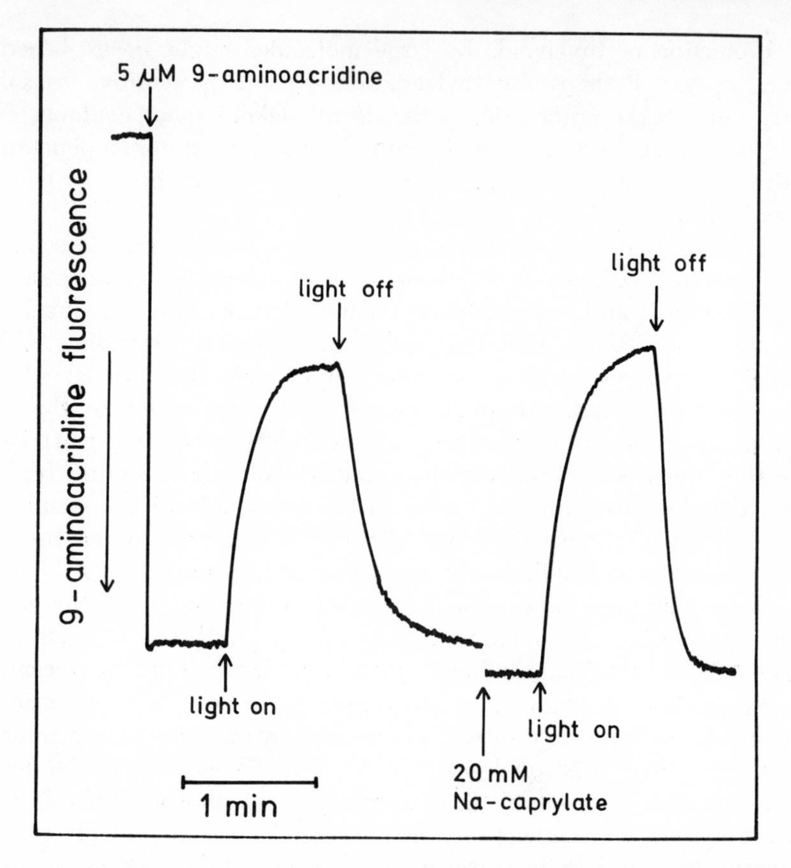

Figure 11. Formation of a trans-thylakoid proton gradient by intact chloroplasts as indicated by the quenching of 9-aminoacridine fluorescence (86, 87), and a subsequent efflux of protons from the thylakoids on darkening. Note accelerated proton efflux in the presence of sodium caprylate. Conditions: Intact spinach chloroplasts were suspended in isotonic sorbitol buffer (20 μg chlorophyll ml⁻¹) and illuminated with saturating red light in the presence of 0.5 mM methylviologen as electron acceptor.

or other semipolar compounds of suitable structure and concentration, proton pumping still produced large proton gradients in the light. However, when the light was turned off, protons leaked out considerably faster than they did in the absence of phenylpyruvate. The available data suggest that salts having an anion that can bind to cationic membrane sites and exert hydrophobic effects will increase the proton conductivity of thylakoids. Light scattering measurements indicate that the permeability to sugars and other compounds is also increased (*51*).

Figure 12 shows preservation of membrane function during freezing of thylakoids in the presence of different concentrations of sodium caprylate or sodium phenylpyruvate. Other solutes present in the system

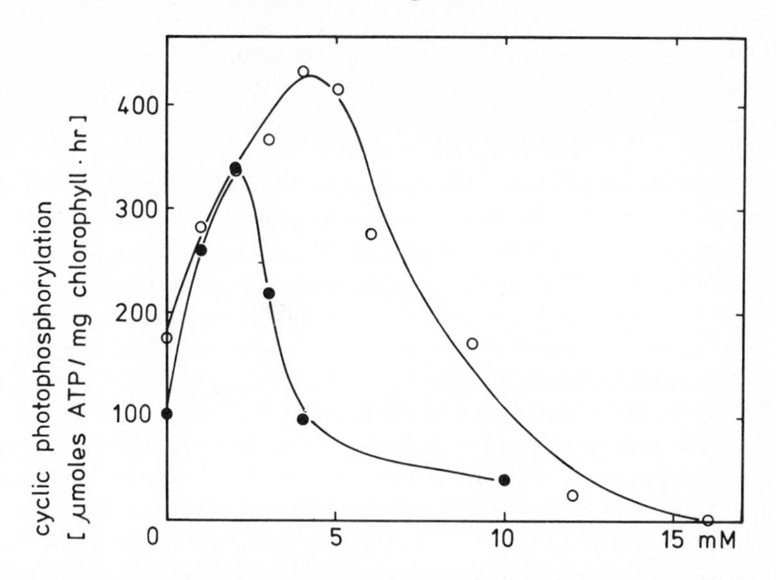

Figure 12. Preservation of thylakoid function after freezing for 3 hours at −25°C in the presence of different concentrations of sodium phenylpyruvate (○) or sodium caprylate (●). Other solutes added to the thylakoid suspension before freezing were, 100 mM NaCl and 120 mM sucrose in the caprylate experiment and 100 mM NaCl and 150 mM sorbitol in the phenylpyruvate experiment.

were sorbitol and NaCl. Their molar ratio was kept constant at a level producing considerable inactivation during freezing to −25°C in the absence of phenylpyruvate or caprylate. When phenylpyruvate or caprylate was added at a very low concentration, a significant increase was observed in the resistance of the membranes to freezing. Protection was optimal in the presence of 2 m*M* caprylate or 4 m*M* phenylpyruvate. Further increases in concentration first decreased protection and finally led to complete membrane inactivation during freezing. When the initial concentration of sorbitol and NaCl was higher than that used for the experiment of Figure 12, a correspondingly higher concentration of sodium phenylpyruvate or caprylate was necessary to achieve maximum protection during freezing. Thus, depending on concentration, phenylpyruvate, and caprylate acted either as cryoprotectants or as cryotoxic solutes. Suitable controls established that cryoprotection by these compounds was not based on colligative action. For instance, the sorbitol concentration had to be raised by about 50 m*M* to get the same protection that was produced by 3 m*M* phenylpyruvate.

Furthermore, protection is not highly specific, since isoleucine and sodium decenylsuccinate protected the membranes to essentially the same extent as caprylate or phenylpyruvate. However, membrane protection in the presence of phenylpyruvate is dependent on the composition of the medium in which the thylakoids are suspended. Protection has been observed in solutions containing balanced concentrations of NaCl and either sorbitol, sucrose, threonine, or sodium succinate (*51, 68*). The permeability of thylakoids to these compounds is low. Phenylpyruvate fails to increase membrane protection in solutions containing NaCl and glycerol or methanol. The latter compounds penetrate thylakoids rapidly. These data suggest that the increased protection seen in the presence of low concentrations of semipolar solutes, such as sodium phenylpyruvate, is the result of increased membrane permeability, which permits leakage of normally nonpenetrating cryoprotectants to the inside of the membranes where protection is also required.

Naturally, the manipulation of membrane permeability is a dangerous matter, since membrane function is intimately related to membrane permeability. Manipulation that goes too far can easily cause membrane damage, as occurs in the presence of increased concentrations of caprylate and phenylpyruvate (Figure 12). High concentrations of these compounds also cause rapid membrane inactivation at 0°C. Still, manipulation of membrane permeability as a means of increasing resistance to freezing appears possible in practice. For example, Kuiper (*88*) reported that sodium decenylsuccinate increased frost hardiness in plants. That other workers have failed to observe protection against freezing damage by decenylsuccinate and instead have reported increased damage (*89, 90*) is not surprising in view of the potential membrane toxicity of compounds capable of altering membrane structure.

Acknowledgment

We are grateful to Mrs. U. Behrend for competent technical assistance and to Prof. O. Fennema and two unknown reviewers for critical and helpful comments. Original research reported in this paper was supported by the Deutsche Forschungsgemeinschaft.

Literature Cited

1. Lovelock, J. E. *Biochim. Biophys. Acta* **1953,** *10,* 414–426.
2. Lovelock, J. E. *Proc. R. Soc. Med.* **1954,** *47,* 60–65.
3. Jagendorf, A. T.; Avron, M. *J. Biol. Chem.* **1958,** *231,* 377–290.
4. Duane, W.; Krogmann, D. W. *Biochim. Biophys. Acta* **1963,** *71,* 195–196.
5. Heber, U.; Santarius, K. A. *Plant Physiol.* **1964,** *39,* 712–719.
6. Santarius, K. A.; Heber, U. *Cryobiology* **1970,** *7,* 71–78.
7. Santarius, K. A. *Biochim. Biophys. Acta* **1973,** *291,* 35–50.
8. Porter, V. S.; Denning, N. P.; Wright, R. C.; Scott, E. M. *J. Biol. Chem.* **1953,** *205,* 883–891.
9. Araki, T. *Cryobiology* **1977,** *14,* 144–150.
10. Farrant, J. *Nature* **1965,** *205,* 1284–1287.
11. Souzu, H. *Arch. Biochem. Biophys.* **1967,** *120,* 344–351.

12. Levitt, J. "Responses of Plants to Environmental Stresses"; Academic: New York, 1972.
13. Meryman, H. T.; Williams, R. J.; Douglas, M. S. J. *Cryobiology* **1977,** *14,* 287–302.
14. Heber, U.; Santarius, K. A. "Temperature and Life"; Precht, H., Christophersen, J., Hensel, H. Larcher, W., Eds.; Springer: Berlin, 1973, pp 232–263.
15. Steponkus, P. L.; Garber, M. P.; Myers, S. P.; Lineberger, R. D. *Cryobiology* **1977,** *14,* 303–321.
16. Asahina, E. In "Cellular Injury and Resistance in Freezing Organisms"; Asahina, E., Ed.; *Low Temp. Sci., Ser. B (Teion Kagaku, Seibutsu-Hen)* **1967,** *II,* 211–229.
17. Mazur, P. *Fed. Proc. Am. Soc. Exp. Biol.* **1965,** *24,* 175–182.
18. Mazur, P. *Science* **1970,** *168,* 939–949.
19. Mazur, P.; Leibo, S. P.; Chu, E. H. J. A. *Exp. Cell Res.* **1972,** *71,* 345–355.
20. Santarius, K. A.; Heber, U. *Planta* **1967,** *73,* 109–137.
21. Santarius, K. A. *Plant Physiol.* **1971,** *48,* 156–162.
22. Heber, U.; Ernst, R. In "Cellular Injury and Resistance in Freezing Organisms"; Asahina, E., Ed.; *Low Temp. Sci., Ser. B (Teion Kagaku, Seibutsu-Hen)* **1967,** *II,* 63–77.
23. Heber, U.; Tyankova, L.; Santarius, K. A. *Biochim. Biophys. Acta* **1971,** *241,* 578–592.
24. Santarius, K. A. *Planta* **1969,** *89,* 23–46.
25. Farrant, J.; Woolgar, A. E. In "The Frozen Cell"; Wolstenholf, G. E. W., O'Connor, M., Eds.; J. & A. Churchill: London, 1970; pp 97–119.
26. Ryrie, I. J.; Jagendorf, A. T. *J. Biol. Chem.* **1971,** *246,* 582–588.
27. Clendenning, K. A. In "Encyclopedia of Plant Physiology"; Ruhland, W., Ed.; Springer: Heidelberg, 1960; Vol. V, 1, pp 736–772.
28. Heber, U.; Tyankova, L.; Santarius, K. A. *Biochim. Biophys. Acta* **1973,** *291,* 23–37.
29. Heber, U. *Cryobiology* **1968,** *5,* 188–201.
30. Heber, U.; Kempfle, M. *Z. Naturforsch. Teil B* **1970,** *25b,* 834–842.
31. Volger, H. G.; Heber, U. *Biochim. Biophys. Acta* **1975,** *412,* 335–349.
32. Santarius, K. A.; Ernst, R. *Planta* **1967,** *73,* 91–108.
33. Santarius, K. A. *Planta* **1973,** *113,* 105–114.
34. Santarius, K. A.; Heber, U. (1972) *Proc. Colloq. on the Winter Hardiness of Cereals, Agric. Res. Inst., Hungarian Academy of Science,* Martonvasar, pp 7–29.
35. Lovelock, J. E. *Biochim. Biophys. Acta* **1953,** *11,* 28–36.
36. Rostand, J. *C.R. Acad. Sci.* **1946,** *234,* 2301–2312.
37. Polge, C.; Smith, A.; Parkes, A. S. *Nature* (London) **1949,** *164,* 666.
38. Meryman, H. T. *Nature (London)* **1968,** *218,* 333–336.
39. Neumann, J.; Jagendorf, A. T. *Arch. Biochem. Biophys.* **1964,** ***107,*** **109–119.**
40. Mitchell, P. *Biol. Rev.* **1966,** *41,* 445–502.
41. Heber, U. *Plant Physiol.* **1967,** *42,* 1343–1350.
42. Garber, M. P.; Steponkus, P. L. *Plant Physiol.* **1976,** *57,* 673–680.
43. Packer, L.; Crofts, A. R. In "Current Topics in Bioenergetics"; Sanadi, D. R., Ed.; Academic: New York, 1967; Vol. 2, pp 24–64.
44. Crofts, A. R.; Deamer, D. W.; Packer, L. *Biochim. Biophys. Acta* **1967,** *131,* 97–118.
45. Deamer, D. W.; Packer, L. *Biochim. Biophys. Acta* **1969,** *172,* 539–545.
46. Karlish, S. J. D.; Shavit, N.; Avron, M. *Eur. J. Biochem.* **1969,** *9,* 291–298.
47. Deamer, D. W.; Crofts, A. R.; Packer, L. *Biochim. Biophys. Acta* **1967,** *13,* 81–96.
48. Volger, H. G.; Heber, U.; Berzborn, R. J. *Biochim. Biophys. Acta* **1978,** *511,* 455–469.

49. Vanderkooi, G. *Ann. N.Y. Acad. Sci.* **1972**,*195*, 6–15.
50. Kamenietzky, A.; Nelson, N. *Plant Physiol.* **1975**, *55*, 282–287.
51. Overbeck, V., Diplomarbeit, University of Düsseldorf, Library no. 415/233 586 (1974).
52. Uribe, E. G.; Jagendorf, A. T. *Arch. Biochem. Biophys.* **1968**, *128*, 351–359.
53. Lien, S.; Berzborn ,R. J.; Racker, E. *J. Biol. Chem.* **1972**, *247*, 3520–3524.
54. Avron, M. *Biochim. Biophys. Acta* **1963**, *77*, 699–702.
55. Strotmann, H.; Hesse, H.; Edelmann, K. *Biochim. Biophys. Acta* **1973**, *314*, 202–210.
56. Howell, S. H.; Moudrianakis, E. N. *Proc. Nat. Acad. Sci. USA.* **1967**, *58*, 1261–1268.
57. Berzborn, J. *Z. Naturforsch., Teil B* **1968**, *23*, 1096–1104.
58. Berzborn, R. J. *Z. Naturforsch., Teil B* **1969**, *24*, 436–446.
59. Racker, E.; Hauska, G. A.; Lien, S.; Berzborn, R. J.; Nelson, N. *Proc. Int. Congr. Photosynth. Res., 2nd, 1971* **1972**, *2*, 1097–1113.
60. Trebst, A. *Annu. Rev. Plant Physiol.* **1974**, *25*, 423–458.
61. Hofmeister, F. *Arch. Exp. Pathol. Pharmakol.* **1888**, *24*, 247–262.
62. Abernethy, J. L. *J. Chem. Educ.* **1967**, *33*, 177–180.
63. Ibid., **1967**, *44*, 364–370.
64. Larsen, J. W.; Magid, L. J. *J. Am. Chem. Soc.* **1974**, *96*, 5774–5782.
65. Taylar, R. P.; Kuntz, I. D. *J. Am. Chem. Soc.* **1972**, *94*, 7963–7965.
66. Heber, U.; Kirk, M. R.; Gimmler, H.; Schäfer, G. *Planta* **1974**, *120*, 31–46.
67. Heber, U., unpublished data.
68. Heber, U.; Santarius, K. A. In "Water and Plant Life"; Lange, O. L., Kappen, L., Schulze, E.-D., Eds.; *Ecol. Stud.* **1976**, *19*, 253–267.
69. Micronescu, S.; Simpson, J. F. *Cryobiology* **1975**, *12*, 581–589.
70. Heber, U. In "The Frozen Cell"; Wolstenholme, G. E. W., O'Connor, M., Eds.; Churchill: London, 1970; pp 175–188.
71. Lovelock, J. E. *Biochem. J.* **1954**, *56*, 265–270.
72. Zade-Oppen, A. M. M. *Experientia* **1970**, *26*, 1087–1088.
73. Woolgar, A. E. *Cryobiology* **1974**, *11*, 44–51.
74. Tumanov, I. I.; Trunova, T. I. *Fiziol. Rast.* **1963**, *10*, 176–182.
75. Tumanov, I. I.; Trunova, T. I.; Smirnova, N. A.; Zvereva, G. N. *Fiziol. Rast.* **1975**, *23*, 132–138.
76. Williams, R. J.; Meryman, H. T. *Plant Physiol.* **1970**, *45*, 752–755.
77. Siminovitch, D.; Singh, J.; de la Roche, I. A. *Cryobiology* **1975**, *12*, 144–153.
79. de Silva, N. S.; Weinberger, P.; Kates, M.; de la Roche, I. A. *Can. J. Bot.* **1975**, *53*, 1899–1905.
79. Willemot, C.; Hope, H. J.; Williams, R. J.; Michaud, R. *Cryobiology* **1977**, *14*, 87–93.
80. Crichley, C. Doctoral Thesis, University of Düsseldorf, 1977.
81. McKenzie, J. S.; Weiser, J.; Stadelmann, E. J.; Burke, M. J. *Plant Physiol.* **1974**, *54*, 173–176.
82. Kacperska-Palacz, A.; Dugokecka, E.; Breitenwald, J.; Wcislinska, B. *Biol. Plant.* **1977**, *19*, 10–17.
83. Kacperska-Palacz, A.; Jasinska, M.; Sobczyk, E. A.; Wcislinska, B. *Biol. Plant.* **1977**, *19*, 18–26.
84. Rochat, E.; Therrien, H. P. *Can. J. Bot.* **1975**, *53*, 536–543.
85. Garber, M. P.; Steponkus, P. L. *Plant Physiol.* **1976**, *57*, 673–680.
86. Schuldiner, S.; Rottenberg, H.; Avron, M. *Eur. J. Biochem.* **1972**, *25*, 64–70.
87. Tillberg, J.; Giersch, C.; Heber, U. *Biochim. Biophys. Acta* **1977**, *461*, 31–47.
88. Kuiper, P. J. C. *Science* **1964**, *146*, 544–546.

89. Newman, E.; Cramer, P. *Plant Physiol.* **1966**, *41*, 606–609.
90. Green, D. G.; Ferguson, W. S.; Warder, F. G. *Plant Physiol.* **1970**, *45*, 1–3.
91. Heber, U.; Purczeld, P. In Photosynthesis 1977"; Hall, D. O., Coombs, J., Goodwin, T. W., Eds.; *Proc. Int. Congr. Photosynth., 4th;* The Biochemical Society: London, 1978; pp 107–118.
92. Mazur, P.; Miller, R. H. *Cryobiology* **1976,** *13*, 523–536.

RECEIVED June 16, 1978.

9

Delocalization of Mitochondrial Enzymes During Freezing and Thawing of Skeletal Muscle

REINER HAMM

German Federal Institute of Meat Research, Kulmbach,
Federal Republic of Germany

The influences of freezing and thawing of bovine and porcine skeletal muscles on the subcellular distribution of the mitochondrial enzymes aconitase (AC), fumarase (FU), malate dehydrogenase (MDH), succinic dehydrogenase (SDH), glutamic dehydrogenase (GLDH), glutamic pyruvic transaminase (GPT), and the mitochondrial isozyme of the glutamic oxaloacetic transaminase (GOT_M) were investigated. Freezing at −5°C had only a slight effect but between −10° and −60° the release of AC, FU, MDH, GLDH, GPT, and GOT_M from mitochondria into the sarcoplasm increased with decreasing temperature, apparently because of increased damage to the mitochondria. SDH was not released. The rate of freezing (between 0.1°/min and 7°/min) had little effect on the release of enzymes. It is suggested that mitochondrial damage is caused mainly by dehydration during ice formation. Release of GOT_M can be reliably used to distinguish between fresh and frozen-thawed red meats, poultry, and carp.

In view of the complicated microstructure of the muscle fiber one would expect that formation of ice crystals within the very small cavities would cause some damage to various cellular elements. A point of particular interest is: "What influence does freezing of muscle tissue have on the sensitive membranes associated with mitochondria, the sarcoplasmic reticulum, lysosomes, and other subcellular particles?" In order

0-8412-0484-5/79/33-180-191$5.00/0

to answer this question, the changes that occur in subcellular structures during freezing must be elucidated. This can be done by electron microscopy or by biochemical methods which sometimes indicate subtle changes that are not visible in electron micrographs. As to the biochemical methods, enzymatic studies are of particular importance. Enzyme proteins are associated with membranes or are located within subcellular organelles. Damage to these organelles or membranes can be recognized by a partial or complete release of enzymes into the sarcoplasmic fluid. If the enzyme released is specific for the subcellular organelle in question, its release provides information about the type and extent of damage of this particular organelle.

Investigations of this type are of importance to medical researchers and physiologists, as well as to food scientists (*1*). For example, enzymes released from cell organelles by freezing and thawing might reach their substrates easier and, therefore, might be more active than in the bound state. Thus, release of dehydrogenases from mitochondria can influence the redox potential of tissue or release of proteolytic enzymes from lysosomes can accelerate the breakdown of muscle proteins (*2*). Furthermore, one might expect that lipids in cell membranes would be more sensitive to autoxidation after membrane damage.

In the field of food science, knowledge about disintegration of cell membranes is important if deterioration of meat quality by freezing and freeze-storage is to be more fully understood (*3*). The presence of membrane-bound enzymes in the muscle-press juice after freezing and thawing could lead to a method for differentiating between nonfrozen meat and frozen and thawed meat.

Enzymes Studied and Their Locations in Mitochondria

In our laboratory we studied the influence of freeze-thawing of muscle on mitochondria, that is, on the release of certain mitochondrial enzymes into the sarcoplasm. The enzymes studied were: aconitase (AC; E.C.4.2.1.3), fumarase (FU; E.C.4.2.1.2), glutamate dehydrogenase (GLDH; E.C.1.4.1.2), malate dehydrogenase (MDH; E.C.1.1.1.37), succinic dehydrogenase (SDH; E.C.1.3.99.1), glutamic oxaloacetic transaminase (GOT; aspartate aminotransferase; E.C.2.6.1.1), glutamic pyruvic transaminase (GPT; alanine aminotransferase; E.C.2.6.1.2), and myokinase (MK; adenylate kinase; E.C.2.7.4.3).

For a correct interpretation of the type of mitochondrial damage it is necessary to know the original sites of enzymes in the mitochondria. One should also realize that mitochondria consist of an outer membrane, an inner membrane, the cristae, the intracristal space, and the mitochondrial matrix. When we started our work, the locations of some enzymes in liver mitochondria had been determined but no information was available concerning the locations of the above-mentioned enzymes in mitochondria of skeletal muscle. Therefore, we investigated the locations of these enzymes in mitochondria isolated from porcine psoas muscle using different treatments such as homogenization with phosphate buffer plus Triton X-100, suspension in distilled water or sucrose–tris–buffer with and without added digitonin, ultrasonic treatment, or freezing and thawing (*1*).

We found that SDH is tightly integrated in the mitochondrial membrane; AC, FU, MDH, and GOT are more or less tightly associated with the mitochondrial membrane; GPT (and MK) are apparently located between the outer and inner membranes (intracristal space); and GLDH is apparently dissolved in the mitochondrial matrix (*1*). Only small amounts of SDH, FU, and AC can be found in the sarcoplasm (present in the press juice from muscle pre or post rigor) of red and white muscles from pigs and cattle (*4*), whereas certain isozymes of GOT (*5*) and MDH (*4*) are located in the mitochondria and other isozymes exist in the sarcoplasm. By means of electrophoresis on cellulose acetate membranes we isolated two GOT isozymes of skeletal muscle, namely the sarcoplasmic isozyme GOT_S and the mitochondrial isozyme GOT_M (*2*). This was the first report that these isozymes existed in porcine and bovine skeletal muscle (*2, 3, 6*).

The release of enzymes from mitochondria by treatment of the tissue was determined by pressing juice from a small piece of muscle and analyzing this juice for enzyme activity. The activity of a given enzyme was then expressed as a percent of the total activity of that enzyme existing in the supernatant fluid of a muscle homogenate prepared using phosphate buffer–Triton X-100 (*6, 8*). Most of the experiments were carried out with white and red muscles of pigs and cattle. The activities of the different enzymes were determined as described by other authors (*4, 7*).

Postmortem Changes in the Subcellular Distribution of Enzymes

Before the influence of freezing was studied, it was necessary to determine whether postmortem (p.m.) storage of muscle at temperatures above freezing (+4°C) influenced the subcellular distribution of the enzymes investigated. As is well known, storage of muscle p.m. results in drastic changes such as development of rigor mortis and a decline in pH from above 7 to about 5.5.

The total activity of the enzymes investigated did not change during development of rigor mortis (*6, 8*). GLDH, which immediately p.m. is located solely in the mitochondria (no activity in the sarcoplasm), is released during storage of the tissue for 48 hours at 4°C (Figure 1). The other mitochondrial enzymes studied are not liberated under such conditions (*8*). The release of GLDH indicates an efflux of the mitochondrial matrix. The mitochondrial membranes, however, seem to remain essentially unchanged. This type of change is probably due to a swelling of mitochondria caused by the decline of pH p.m. (*8*).

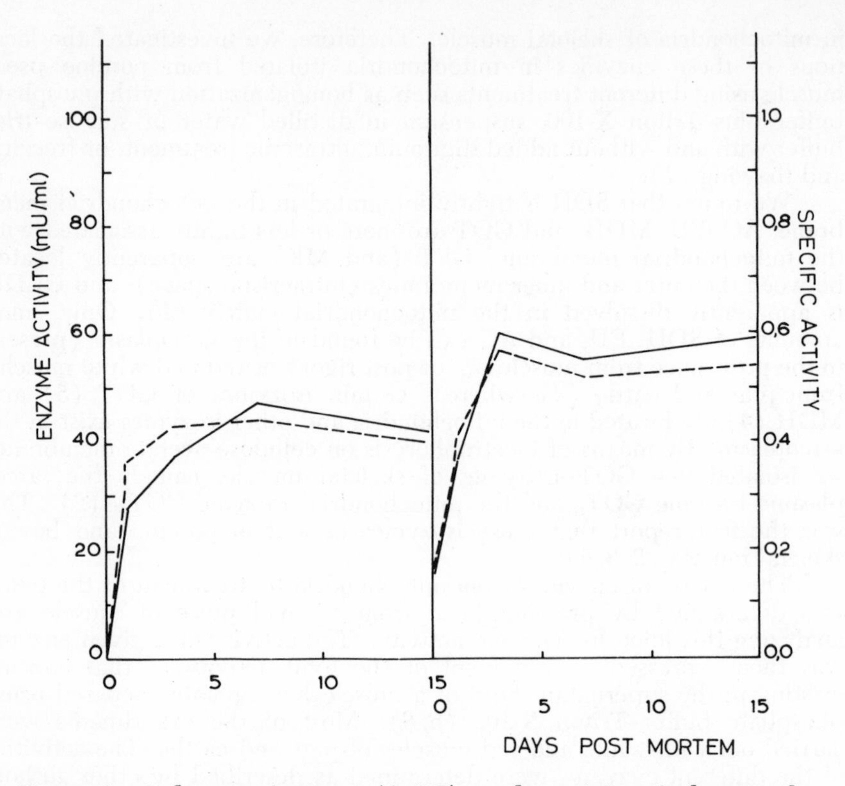

Figure 1. Influence of storage (+4°C) on the activity of glutamic dehydrogenase in muscle press juice from bovine longissimus dorsi (time postmortem) (● — ●) absolute activity (mU/ml); (○ - - - ○) specific activity. Two animals (8).

Influence of Freezing and Thawing on the Subcellular Distribution of Enzymes

Freezing the tissue at −20°C before or after rigor mortis (*9, 10*), and frozen-storage for 48 hours (*11*) did not significantly change the total activity of the enzymes investigated. Freezing and thawing of prerigor muscle, however, cause a highly significant release of AC, FU, GLDH, and MDH from the mitochondria into the sarcoplasm (Table I) (*9, 10*). Freezing of muscle after development of rigor mortis has the same effect (Table II) (*10, 12*), except GLDH, which is liberated during rigor without freezing, then is inactivated partially by freezing and thawing (*10*). Freezing and thawing also cause a considerable release of mitochondrial GOT (Figure 2; Table III) (*9, 12*). The same is true for MK (*13*).

Table I. Enzyme Activities[a] in the Press Juice of Prerigor Bovine Longissimus Muscle before and after Freezing[b] (*10*)

Enzyme	*Animal No.* I		II		III	
	Before (%)	*After (%)*	*Before (%)*	*After (%)*	*Before (%)*	*After (%)*
AC	13.3	41.7	12.9	33.9	11.9	31.0
FU	6.7	14.6	6.4	13.5	11.0	13.8
GLDH	15.3	101.3	0.0	58.7	10.5	75.5
MDH	72.8	74.3	—	—	54.9	63.7
SDH	0.8	0.0	1.1	3.2	0.0	2.4
GPT	52.0	81.1	—	—	—	—

[a] Percent of the total extractable activity.
[b] Freezing at −20°C.

Table II. Enzyme Activities[a] in the Press Juice of Postrigor Bovine Longissimus Muscle before and after Freezing[b] (*10*)

Enzyme	*Animal No.* I		II		III	
	Before (%)	*After (%)*	*Before (%)*	*After (%)*	*Before (%)*	*After (%)*
AC	10.4	42.5	7.6	27.4	3.8	17.1
FU	1.6	25.3	2.0	12.9	14.0	25.9
GLDH	76.6	44.9	95.6	30.0	100.0	62.5
MDH	77.7	98.0	—	—	76.8	100.0
SDH	5.3	2.7	4.7	1.6	3.3	0.0

[a] Percent of total extractable activity.
[b] Freezing at −20°C.

Table III. Influence of Freezing[a] and Thawing of Longissimus Muscle (Postrigor) on the Activities of GOT Isozymes in the Muscle Press Juice[b] (*9*)

Species	*Treatment of Tissue*	*Number of Animals*	GOT_S % $\bar{x}$	GOT_S % s	GOT_M % $\bar{x}$	GOT_M % s
Cattle	unfrozen	15	98.0	1.83	2.0	1.83
Cattle	frozen	14	71.9	6.43	28.1	6.46
Pig	unfrozen	10	94.0	4.81	5.0	2.47
Pig	frozen	10	64.9	8.26	35.1	8.26

[a] Freezing at −20°C.
[b] Percent of total GOT activity in the muscle press juice.

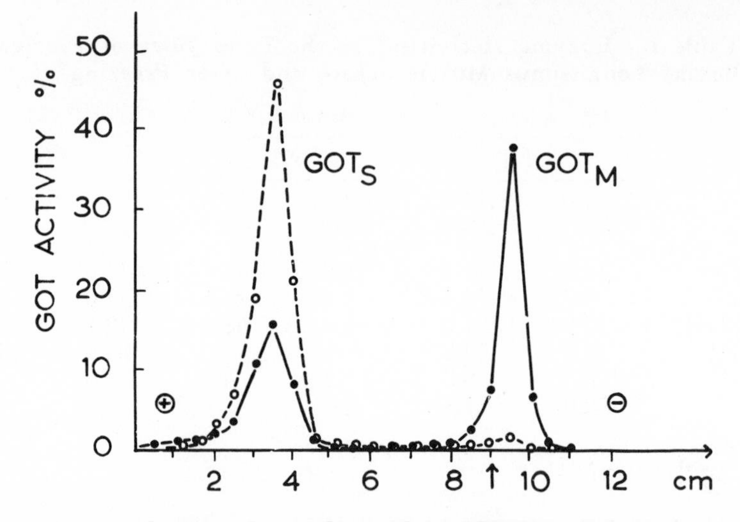

Figure 2. Results from low voltage electrophoresis of the press juice from nonfrozen and frozen-thawed porcine muscle. Psoas muscle, excised 5 days postmortem and frozen at −20°C for 20 days. (Ordinate) GOT activity as percent of total GOT activity; (arrow) start line (9). (○ - - - ○) Nonfrozen; (● — ●) frozen and thawed.

Repeated freezing and thawing of bovine muscle increase release of GOT_M (*9*) and AC (*12*) but not FU (Table IV). It should be mentioned that the increased enzyme activity in muscle press juice after freezing and thawing is not due to freeze-activation of enzymes already present in the sarcoplasm, because freeze-thawing of muscle-press juice does not cause a significant increase in the activities of AC, FU, GLDH, MDH, and SDH (*10*). From these results we can conclude that freezing at −20°C and thawing cause considerable damage to mitochondrial membranes because enzymes attached to the membranes are released. A complete disintegration of the membranes, however, does not seem to occur because SDH is not delocalized (Tables I and II) (*10, 12*).

Table IV. Influence of Repeated Freezing[a] and Thawing of Bovine Longissimus Muscle (Postrigor) on the Enzyme Activities in Muscle-Press Juice[b] (*9, 10*)

Treatment of Muscle Tissue	*AC (%)*	*FU (%)*	*GOT_M (%)*
Unfrozen	7.3	5.9	2.8
Once frozen	20.7	42.8	19.8
Twice frozen	34.6	28.7	25.9
Three times frozen	51.6	14.7	27.7

[a] Freezing at −20°C.
[b] Percent of total extractable activity.

Frozen storage of muscle tissue for three months at −20°C does not cause an additional increase in enzyme activities in the muscle-press juice obtained after thawing (*10*), except for GOT_M (*9*). However, it is possible that a further release of mitochondrial enzymes does occur along with a compensating inactivation of enzymes.

Influence of Temperature and Rate of Freezing on Enzyme Activity

Freezing conditions have a remarkable influence on the release of mitochondrial enzymes. Samples of bovine longissimus muscle (post rigor) were frozen to −5°, −10°, −20°, −40°, −60° and −80°C at "slow" and "fast" cooling rates (Table V), then subsequently thawed at room temperature. For freezing, 8.5 × 5.0 × 2.5 cm samples were sealed in polyethylene bags and immersed in a methanol bath. In the "fast"-freezing experiments, the bath was maintained at the temperature indicated in Figure 3 and the samples were kept in the bath until their centers reached the temperatures indicated. In the "slow"-freezing experiments, samples were immersed in the bath at a bath temperature of +15°C, then the bath temperature was lowered at a rate of 0.2°C per minute until the desired final temperature was attained (*12*). At high subfreezing temperatures (−3° to −8°C) very little release of mitochondrial enzymes occurred; between −8°C and −60°C, however, the release of AC, FU, GOT_M, and GPT increased with falling temperature (Figure 3). A further decrease in the final temperature did not seem to cause additional release of enzymes from the mitochondria (Figure 3) (*12*).

Contrary to the effect of freezing temperature, the rate of freezing did not have a significant influence on mitochondrial damage as evidenced by the almost equal release of mitochondrial enzymes (FU, AC, GOT_M,

Table V. Rate of Temperature Decrease Between 0° and −5°C during "Slow" and "Fast" Cooling[a] (*12*)

Final Temperature in the Center of the Sample (°C)	*"Slow" Freezing (°C/min)*	*"Fast" Freezing (°C/min)*
−5	0.058	0.072
−10	0.097	0.222
−20	0.097	0.497
−40	0.106	2.220
−60	0.107	4.010
−80	0.097	7.150

[a] The experimental conditions are described in the text.

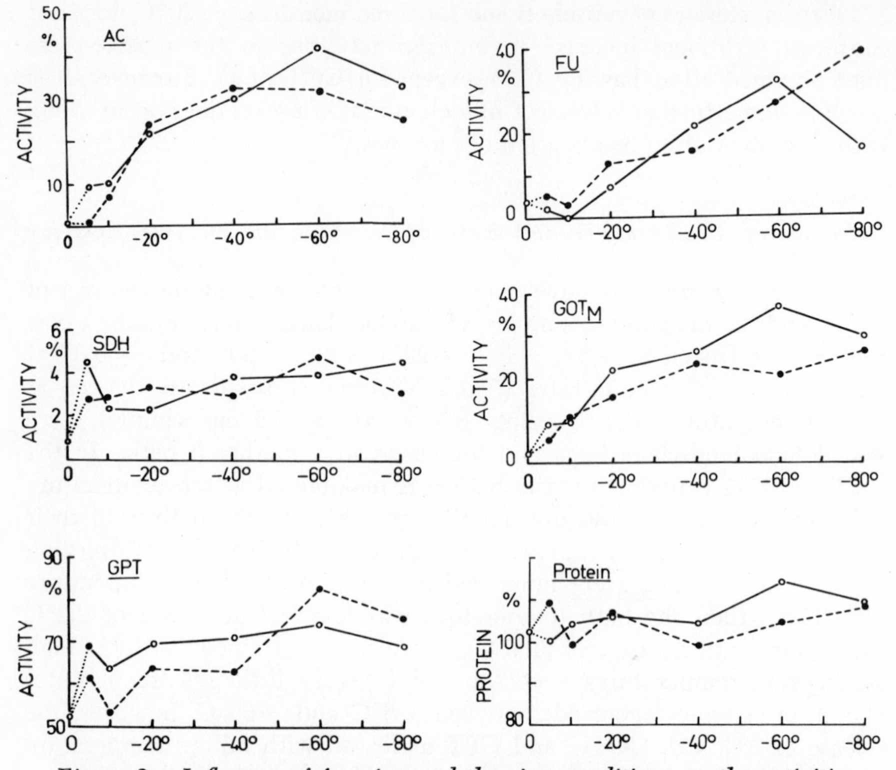

Figure 3. Influence of freezing and thawing conditions on the activities of aconitase (AC), fumarase (FU), succinic dehydrogenase (SDH), mitochondrial glutamic oxaloacetic transaminase (GOT_M), and glutamic pyruvic transaminase (GPT) in the muscle press juice from bovine longissimus dorsi (postrigor). Percent of the total extractable activity. (Ordinate) Ultimate sample temperature achieved. Average of muscles from 3 animals. (● - - - ●) "slow" freezing; (○ — ○) "fast" freezing. Point "0" of the temperature scale means "nonfrozen" (+5°C) (12).

GPT) during "slow"- and "fast"-freezing (Figure 3) (*12*). This is somewhat surprising since "slow"-freezing results in the formation of larger ice crystals than "fast"-freezing (*14, 15, 16*). However, factors other than ice crystal size may be of overriding importance (*14, 16*).

Causes of Damage to Mitochondria during Freezing and Thawing

From these results it can be concluded that damage to mitochondrial membranes, as observed in this study, was more closely related to the extent of ice formation than to the size of ice crystals. Therefore, this damage might not be caused by mechanical injury of membranes but rather by removal of water from the membranes during freezing or, in other words, by a dehydration process. Water is necessary for the stability of membrane structures. Removal of bound water by freezing has a detrimental effect, weakening the lattice structures of water and promoting denaturation and subsequent breakdown of the membranes. This explanation of freezing damage is in agreement with ideas already stated by Luyet and Grell (*17*) and by Karow and Webb (*18*), and later supported by many other authors (*10, 12, 19, 20, 21, 22*). It should be pointed out that at −5°C about 80 percent of the freezable water in muscle is frozen (*15, 16*) and that damage to mitochondria, as evidenced by release of enzymes, occurs during freezing of the remaining 20 percent.

Freeze-damage to tissue is often attributed to an increased ion concentration in the unfrozen part of the cell water (*14, 16, 23, 24*). However, this factor might not play a dominant role in the results observed here. During freezing, superimposition of the effect of increasing ion concentration on the effect of decreasing water activity is the reason that many reactions that are facilitated by ions show a maximum rate at about −5°C (*25*). So, damage of proteins during frozen storage is maximal around this particular temperature. According to our results, however, damage to mitochondria is minimal at −5°C and increases with declining temperature (Figure 3).

Our finding that releasable enzymes are almost completely released between −40° and −60°C (Figure 3) (*12*) probably can be explained by the fact that the final eutectic temperature of muscle tissue lies in this temperature range (*14, 15*). Below the final eutectic temperature no further ice formation will occur, provided solid–liquid equilibrium has been attained.

With regard to the role that membrane dehydration has in the release of mitochondrial enzymes during the freeze–thaw process, it is interesting that freeze-dehydration of muscle tissue causes a significantly greater release of AC, FU (Figure 4), GLDH (*26*), and GOT_M (Table VI) than the freezing (or freeze-thaw) process alone (*26*).

It should be mentioned that freeze damage to the membranes of cell organelles, such as mitochondria, is probably caused by a combination of factors including dehydration, concentration of electrolytes, and mechanical effects of ice crystals. But in my opinion the dehydration effect is predominant under the conditions used in these experiments.

If freezing of tissues occurs slowly, ice forms in extracellular areas as water flows out of the cell by exosmosis. As a result, the cell dehydrates and does not freeze intracellularly. However, if the cell is cooled rapidly, it cannot lose water fast enough to maintain equilibrium with its environment, and it therefore becomes increasingly supercooled and eventually freezes intracellularly (*27*). Mazur (*27, 28*) suggested that injury from intracellular ice and its subsequent growth by recrystallization is a direct

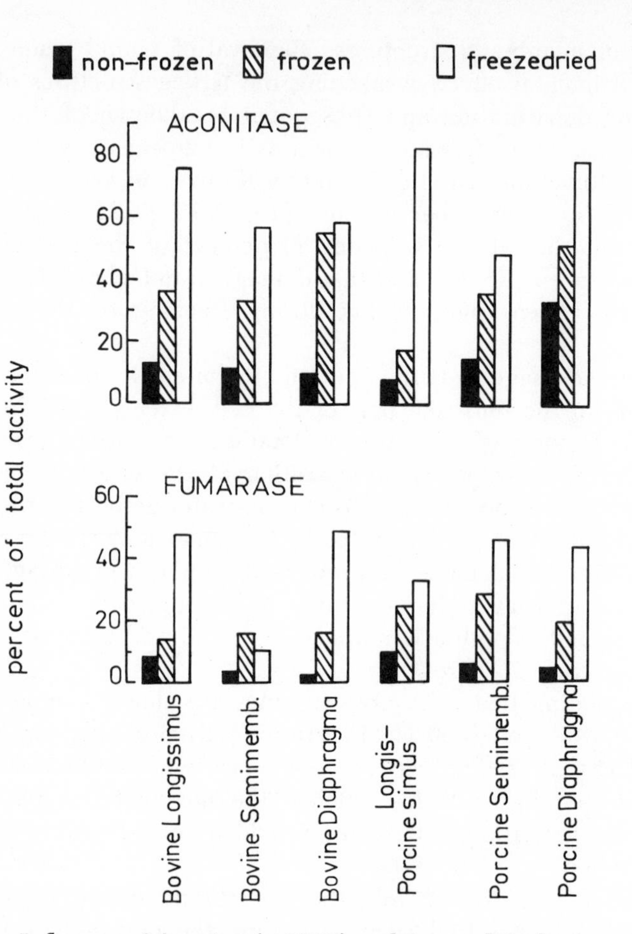

Figure 4. Influence of freezing (−20°C) and freeze-dehydration of pre-rigor muscle tissue on the activities of aconitase and fumarase in the muscle press juice. Percent of the total extractable activity. The press juice was prepared from the thawed tissue and the rehydrated tissue respectively (26).

physical consequence of this ice. The latter type of injury, however, probably was not a significant factor in our experiments because even the "fast"-freezing used might have been "slow"-freezing from a cryobiological point of view, although we did not determine the existence of intracellular ice formation. Greater freezing rates might be of practical interest for freezing small specimens such as organelles, single cells, or single muscle fibers but they are not feasible for freezing whole organs or large pieces of muscle.

Under certain conditions the rate of thawing might influence the type of mitochondrial damage (*29, 30*). In our experiments, all samples were exposed to about the same temperature during thawing but the rate

Table VI. Influence of Freezing[a] and Freeze-Dehydration of Bovine Longissimus Muscle (Postrigor) on GOT and GOT_M in the Muscle-Press Juice[b] (26)

Treatment of Muscle Tissue	*GOT Activity (U/ml)[c]*	*GOT_M Activity as a Percent of Total Extractable Activity*
Fresh	2530	2
Frozen	3146	16
Freeze-dehydrated	4449	35

[a] Freezing at −20°C.
[b] The press juice was prepared from thawed tissue and rehydrated tissue, respectively.
[c] U = units of enzyme activity.

of thawing was dependent on the original temperatures of the frozen samples. As is well known, a tissue remains in the range between 0° and −5°C much longer during thawing than during freezing (*31*). In this temperature range, however, the damage to mitochondria is apparently very small. Thus, thawing conditions would appear to have had little bearing on the results reported here.

The release of mitochondrial enzymes by freezing, described in this paper, is closely related to the phenomenon of latency which is defined as an increase in the activity of mitochondrial or lysosomal enzymes caused by freezing and thawing (*32*). With respect to mitochondria, the enzymes AC, GLDH, MDH, hydroxybutyrate dehydrogenase, isocitrate dehydrogenase, and cytochrome oxidase have been observed to exhibit latent behavior. According to Bendall and de Duve (*33*) the freeze-induced increase in activity of GLDH and MDH should be due to a release of these enzymes from their binding sites in mitochondria. In this context it is of interest that freezing and thawing of muscular granules leads to the release of cathepsin and other acidic hydrolases from lysosomes (*34, 35*).

In studies involving isolated liver mitochondria, release of GLDH, 3-hydroxybutyrate dehydrogenase (*36, 37*), and MK (*38*) by freezing and thawing was demonstrated. Lusena (*36, 37*) concluded from these results that freeze-damage to liver mitochondria occurs in two steps: one rapid step involving damage to the mitochondrial membranes and a second step which occurs more slowly and results in a release of matrix enzymes. At very rapid freezing rates (several hundred °C per minute), which were not used in our experiments, mitochondria undergo a loss of oxidative activity (*39, 40*).

Changes in mitochondrial structures, caused by freezing or frozen storage of muscle (*41, 42*) or liver tissue (*43*), also can be demonstrated by electron microscopy. Such structural changes increase progressively with increases in cooling rate (*39*).

Differentiation Between Fresh and Frozen-Thawed Meat

The release of mitochondrial enzymes by freezing and thawing can be used to differentiate fresh meat and frozen-thawed meat in practical situations. For this purpose the enzyme chosen must fulfill three requirements: (a) it should be released by freezing and thawing but not by aging of meat; (b) its total activity should not decrease markedly during storage of muscle either fresh or frozen; and (c) it should be easily detectable in the muscle-press juice. Among all of the enzymes we have investigated only GOT_M meets these criteria. As already mentioned, GOT_M can be separated from GOT_S by simple electrophoresis on cellulose acetate membranes.

All reagents needed to separate GOT isozymes by electrophoresis are commercially available. After spraying the wet membrane with the Boehringer GOT test solution, the isozymes appear under a UV lamp as dark bands (NAD^+) on a bright fluorescent background (NADH). Electrophoresis of press juice from unfrozen meat results in only one band (GOT_S), whereas press juice from frozen-thawed meat results in two bands (GOT_S and GOT_M). This method is applicable not only to beef and pork (*9, 44*) but also to poultry meat, e.g. meat from chicken, turkey, geese and ducks (Figure 5) (*45*). Two GOT isozymes were present in the press juice obtained from frozen-thawed samples of both breast and thigh muscles of poultry (*46*). Also a differentiation between fresh and frozen-thawed liver is possible using this technique (*47*).

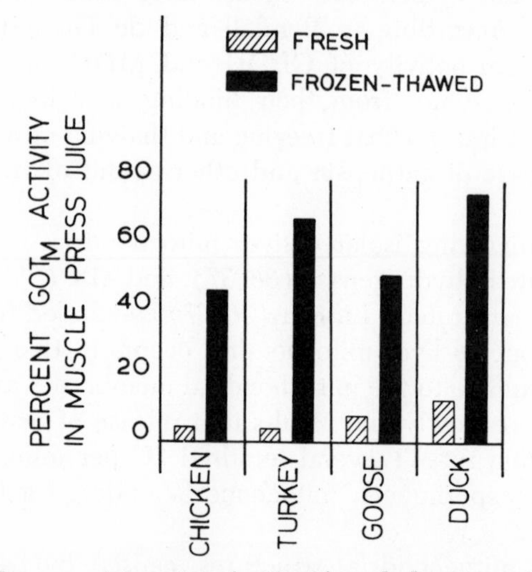

Figure 5. Influence of freezing (−20°C) and thawing of poultry thigh muscles (postrigor) on the GOT_M activity in the muscle press juice. (Ordinate) GOT_M activity as a percent of the total GOT activity in the press juice (45).

Table VII. Influence of Freezing[a] and Thawing of White Carp Muscle (Filet) on the Activities of GOT Isozymes in the Muscle-Press Juice[b] (*49*)

	GOT_S		GOT_{M1}		GOT_{M2}		GOT_{M3}	
	Fresh	*Frozen*	*Fresh*	*Frozen*	*Fresh*	*Frozen*	*Fresh*	*Frozen*
$\bar{x}$	79	16	5	20	8	35	8	30
s	11	8	3	3	5	3	4	7

[a] Freezing at −20°C.
[b] Percent of total GOT activity in the press juice. Average of 10 muscles.

This procedure also can be used to differentiate between fresh and frozen-thawed fish. In fresh, homogenized carp muscle we found four GOT isozymes—one sarcoplasmic and three mitochondrial (*48*). Upon freezing of whole carp, the mitochondrial isozymes (GOT_{M1}, GOT_{M2}, GOT_{M3}) are released (Table VII) (*49*). Thus, press juice from fresh carp muscle exhibits one band, whereas after freezing and thawing four bands appear.

Acknowledgment

I thank Professor Dr. O. R. Fennema, University of Wisconsin, Madison, for valuable discussions and suggestions.

Abbreviations Used

AC = aconitase
FU = fumarase
GLDH = glutamate dehydrogenase
MDH = malate dehydrogenase
SDH = succinic dehydrogenase
GOT = glutamic oxaloacetic transaminase (aspartate aminotransferase)
GOT_S = sarcoplasmic GOT isozyme
GOT_M = mitochondrial GOT isozyme
GPT = glutamic pyruvic transaminase (alanine aminotransferase)
MK = myokinase (adenylate kinase)
p.m. = post mortem
U = units of enzyme activity

Literature Cited

1. Hamm, R.; El-Badawi, A. A.; Tetzlaff, L. Z. *Lebensm.-Unters.-Forsch.* **1972,** *149,* 7.
2. Körmendy, L.; Gantner, G.; Hamm, R. *Biochem. Z.* **1965,** *342,* 31.
3. Körmendy, L.; Gantner, G.; Hamm, R. *Naturwissenschaften* **1965,** *52,* 209.
4. El-Badawi, A. A.; Hamm, R. *Z. Lebensm.-Unters.-Forsch.* **1972,** *149,* 87.
5. Hamm, R. *J. Food Sci.* **1969,** *34,* 449.

6. Hamm, R.; Körmendy, L., Gantner, G. *J. Food Sci.* **1969,** *34,* 446.
7. Gantner, G.; Hamm, R. *Z. Lebensm.-Unters.-Forsch.* **1964,** *126,* 1.
8. El-Badawi, A. A.; Hamm, R. *Z. Lebensm.-Unters.-Forsch.* **1972,***149,* 257.
9. Hamm, R.; Körmendy, L. *J. Food Sci.* **1969,** *34,* 452.
10. Hamm, R.; El-Badawi, A. A. *Z. Lebensm.-Unters.-Forsch.* **1972,** *150,* 12.
11. Hamm, R.; El-Badawi, A. A., unpublished data.
12. El-Badawi, A. A.; Hamm, R. *Z. Lebensm.-Unters.-Forsch.* **1976,** *162,* 217.
13. Hamm, R., Masic, D. D., unpublished data
14. Fennema, O. R. In "Low Temperature Preservation of Foods and Living Matters," Fennema, O. R., Powrie, W. D., Marth, E. H., Eds.; M. Dekker: New York, 1973.
15. Riedel, L. *Kaeltetechnik* **1961,** *13,* 122.
16. Love, R. M. In "Cryobiology," Meryman, H. T., Ed.; Academic, New York, 1966; p 317.
17. Luyet, B. J.; Grell, S. M. *Bio-dynamics* **1936,** *23,* 1.
18. Karow, A. M.; Webb, W. R. *Cryobiology* **1975,** *2,* 99.
19. Meryman, H. T. *Cryobiology* **1971,** *8,* 489.
20. Litvan, G. G. *Cryobiology* **1972,** *9,* 182.
21. Baust, J. G. *Cryobiology* **1973,** *10,* 197.
22. Mazur, P. *Science* **1970,** *168,* 939.
23. Fennema, O. R. *Cryobiology* **1966,** *3,* 197.
24. Meryman, H. T. *Cryobiology* **1971,** *8,* 382.
25. Riedel, L. In "Lebensmittelkonservierung," *DECHEMA-Monogr.* **1969,** *63,* 118.
26. El-Badawi, A. A., Hamm, R., unpublished data.
27. Mazur, P. *Cryobiology* **1977,** *14,* 251.
28. Mazur, P. In "Cryobiology," Meryman, H. T., Ed.; Academic: New York, 1966; p 213.
29. Sherman, J. K. *Cryobiology* **1972,** *9,* 112.
30. Bowers, W. D.; Hubbard, R. C.; Daum, R. C.; Ashbaugh, P.; Nilson, E. *Cryobiology* **1973,** *10,* 9.
31. Luyet, B. J. In "Low Temperature Biology of Foodstuffs," Hawthorn, J., Rolfe, E. J., Eds.; Pergamon: Oxford, 1968.
32. Tappel, A. L. In "Cryobiology," Meryman, H. T., Ed.; Academic, New York, 1966.
33. Bendall, B. S.; DeDuve, C. *Biochem. J.* **1960,** *74,* 444.
34. Parrish, F. C.; Bailey, M. E. *J. Agric. Food Chem.* **1967,** *15,* 88.
35. Stagni, N.; de Bernard, B. *Biochim. Biophys. Acta* **1968,** *170,* 129.
36. Lusena, C. V. *Can. J. Biochem.* **1965,** *43,* 1787.
37. Lusena, C. V.; Dass, C. M. S. *Can. J. Biochem.* **1966,** *44,* 775.
38. Brdiczka, D.; Pette, D.; Brunner ,G.; Miller, F. *Eur. J. Biochem.* **1968,** *5,* 294.
39. Fishbein, W. N.; Griffin, L. *Cryobiology* **1976,** *13,* 542.
40. Araki, T. *Cryobiology* **1977,** *14,* 144.
41. Partmann, W. *Kaeltetechnik* **1964,** *16,* 34.
42. Partmann, W. *Fleischwirtschaft* **1968,** *48,* 1317.
43. Trump, B. F.; Douglas, E. Y.; Arnold, E. A.; Stowell, R. E. *Fed. Proc., Fed. Am. Soc. Exp. Biol.* **1965,** *24,* 144.
44. Hamm, R.; Körmendy, L. *Fleischwirtschaft* **1966,** *46,* 615.
45. Hamm, R.; Masic, D.; Tetzlaff, L. *Z. Lebensm.-Unters.-Forsch.* **1971,** *147,* 71.
46. Masic, D.; Hamm, R.; Tetzlaff, L. *Z. Lebensm.-Unters.-Forsch.* **1971,** *146,* 205.
47. Hamm, R.; Masic, D. *Fleischwirtschaft* **1975,** *55,* 242.
48. Masic, D.; Hamm, R. *Arch. Fischereiwiss.* **1971,** *22,* 110.
49. Hamm, R.; Masic, D. *Arch. Fischereiwiss.* **1971,** *22,* 121.

Received June 16, 1978.

10

Denaturation of Fish Muscle Proteins During Frozen Storage

JUICHIRO J. MATSUMOTO

Department of Chemistry, Sophia University, Kioi-cho 7, Chiyoda-ku, Tokyo, Japan 102

Studies on the freeze denaturation of fish muscle proteins were reviewed with emphasis given to changes in their physico–chemical and biochemical properties during frozen storage. Denaturation of actomyosin commonly occurs during frozen storage and the side-to-side aggregation of myosin molecules appears to play a major role in this reaction. The author's group performed freezing studies with isolated preparations of proteins from fish muscle, i.e., actomyosin, myosin, H–meromyosin (HMM), L–meromyosin (LMM), and actin. Freeze denaturation occurred with individual proteins as well as with their subunits. Not only aggregation but also some conformational changes were observed. Denaturation was inhibited significantly in the presence of added monosodium glutamate (MSG). About 30 compounds were found to inhibit denaturation and their mechanisms of action are discussed.

Changes in the Sensory Attributes of Fish Muscle

Studies of protein denaturation in fish muscle during frozen storage have been carried out to gain scientific knowledge and to provide a basis for supplying foods of better quality. Early studies with fish demonstrated that frozen storage is an excellent means of preventing putrefaction and autolysis of this perishable commodity. However, it was soon learned that frozen fish deteriorates more rapidly than frozen bovine muscle. Frozen fish can exhibit several kinds of quality deterioration depending on the state at which it is examined. For example, a fish that

0-8412-0484-5/79/33-180-205$5.00/0

Table I. Proteins

Group	*Proteins*
Sarcoplasmic proteins (albumins)	myogen
Myofibrillar proteins (globulins)	(actomyosin) myosin H–meromyosin L–meromyosin
	actin tropomyosin troponin, etc.
Stroma (scleroproteins)	collagen elastin

[a] Values from many references were referred to and summarized.

has been frozen for a long period then thawed often exhibits a softened texture, considerable loss of fluid, and changes in flavor and odor. When this same fish is used as a processing material it may differ from fresh fish in terms of texture, water-holding capacity, binding properties, and gelling strength. Finally, it may differ from fresh-cooked fish with respect to texture (toughness, coarseness, dryness, etc.), fluid losses, flavor, and odor.

Many of the changes in frozen fish muscle are attributable to denaturation of proteins and studies of this occurrence will be reviewed here. Several earlier reviews of protein denaturation in frozen fish have appeared in the literature (*1–8*).

Structure and Constituents of Fish Muscle

Although fish muscle is characterized by the presence of dark muscle and myocommata (*9*), most works so far published concentrate on the normal or white muscle because it comprises the major part of the muscle.

The fibers of both white and dark muscles consist of bundles of striated myofibrils each containing thin and thick filaments and various subcellular structures such as the sarcoplasmic reticulum and other organelles (*9, 10, 11*). The size and shape of fish muscle fibers and myofibrils differ somewhat from the corresponding components of mammalian muscles (*9, 11*).

The principal constituents of fish muscle are: water, 66-84%; protein, 15-24%; lipid, 0.1-22%; and mineral substances, 0.8-2% (*12*). Proteins, the major constituent of the dry matter, can be classified into three groups based on solubility. This classification is shown in Table I (*1, 11*). The

of Fish Muscle[a]

Localization	Function	Amount per Total Proteins
cell plasma	glycolytic enzymes	18–30%
		65–80%
thick filaments	contraction	
thin filaments	contraction regulation regulation	
myocommata and cell membranes	connective tissues	3–5%

amount of stroma proteins is much less in fish muscle (3%) than it is in mammalian skeletal muscles (15%). Myofibrillar proteins are the major proteins in fish muscle and they are present in larger amounts than in mammalian skeletal muscle (57–68%).

The size and shape of myofibrillar proteins of fish are shown in Table II. Properties of fish proteins including amino acid composition are generally comparable with those of rabbit skeletal muscle (*11, 13, 14, 15*). However, the myofibrillar proteins of fish differ from those of rabbit in some respects: 1) fish actins go readily into solution in a Guba–Straub medium $\mu = 0.5$ and pH $\sim$ 6.5) to form a viscous actomyosin solution that hinders isolation of pure myosin (*11, 13*) and 2) fish myosins are much more labile than mammalian myosins when stored at nonfreezing temperatures (*16*), or when exposed to proteases or denaturants such as urea and guanidine–HCl (*17*). In the former case (nonfreeze storage), spontaneous side-to-side aggregation occurs with little or no change in conformation (*16*) whereas in the latter case (urea and guanidine–HCl)

Table II. Size and Shape of Fish Myofibril Proteins[a]

Protein	Mol Wt	Sedimentation Coefficient	Intrinsic Viscosity $dL \times g^{-1}$	Shape
Myosin	500,000	6.4 S	1.8–2.3	fibrillar
Actin				
(G-form)	43,000	3.3 S	0.1–0.4	globular
(F-form)		large		fibrillar
Tropomyosin	68,000	2.85 S	0.36	rod
Troponin	80,000			globular

[a] Values from many references were referred to and summarized.

unfolding of peptide chains occurs (*17*). Furthermore, aggregation is frequently encountered with fish actomyosins. These instabilities of fish proteins clearly are related to the ease with which fish protein denatures during frozen storage.

Early Investigations

Denaturation of proteins as related to quality deterioration of frozen-stored fish muscle was first studied by Finn (*18*). Reay (*19*) revealed that during frozen storage of haddock, globulins (salt-soluble proteins) became insoluble whereas albumins (water-soluble proteins) were not affected. Dyer (*20*) demonstrated the relationship between the freeze deterioration of organoleptic properties and denaturation of actomyosin. He found a significant correlation between favorable taste-panel scores and the amount of actomyosin extractable from frozen-stored fillets of cod, halibut (*20*), plaice (*21*), and rosefish (*22*). Sarcoplasmic proteins (water-soluble proteins) underwent no appreciable change during frozen storage.

Dyer's outstanding work stimulated many studies of a similar nature leading to the firm conclusion that denaturation of actomyosin occurs at a significant rate during frozen storage of fish muscle and that the rate of denaturation can be used to estimate the rate of quality change of the fish.

Most of the early work on this subject was done by extracting acto-myosin from fish fillets following frozen storage and thawing. On the other hand, many studies designed to determine the mechanisms of denaturation have involved frozen solutions or suspensions of isolated protein preparations.

Factors Affecting the Rate of Denaturation

Rate of Freezing. It is widely believed that rapid freezing generally results in less denaturation than slow freezing. However, Love studied cod fillets and found that some intermediate rates of freezing resulted in more detrimental textural changes than slow freezing (*23, 24, 25, 26*). The rapidly frozen fillets contained small intracellular ice crystals and exhibited little change in cell structure. The slowly frozen fillets contained large intercellular ice crystals with muscle fibers that were shrunken and tightly grouped. Fillets frozen at intermediate rates contained large intracellular ice crystals that had grown so large that they damaged the cell membranes. Actomyosin was the least soluble in the last instance (*26*). Similar observations were reported by Tanaka (*27*).

Rates of denaturation of muscle proteins in Alaska pollack, as influenced by freezing rate and storage temperature, were studied by Matsuda

(*28*) and by Suzuki et al. (*29*). Muscle frozen in liquid nitrogen contained small intracellular ice crystals and less denaturation of actomyosin than muscle frozen in air at −20°C. Contrary to this finding Love reported that freezing of cod muscle in liquid air caused tightening of texture, a coalescence of myofibrils, and a decrease in extractable proteins (*30, 31*).

Temperature and Duration of Storage. As shown by the works of Dyer (*20, 21, 22*) and Connell (*32*) lowering the storage temperature decreases the rate of protein denaturation. Love's study, involving freezing in liquid air, yielded results of a contrary nature (*30*). Furthermore, Snow (*33*) dissolved cod actomyosin in various inorganic salt solutions and found that freezing to temperatures below the eutectic points of the respective salts denatured the protein, apparently because of a dehydrating effect. However, this result is contradictory with the general observation that proteins of Alaska pollack and sea bass undergo little denaturation when the fillets were stored at −20°C after freezing in liquid nitrogen (*29*).

During frozen storage two events occur simultaneously—denaturation of actomyosin and an increase in the average size of ice crystals (*34, 35*). Moreover, a fluctuating storage temperature accelerates the decline in sensory score, the decrease in the solubility of actomyosin, and the growth of ice crystals (*36, 37*). Thus, the latter occurrence may be related to the first two. In general, refreezing has a detrimental effect on denaturation of proteins in fish muscle (*38, 39, 40*).

Supercooling (*41*) and partial freezing (*42*) which are accompanied by no or little ice formation at a temperature slightly below the freezing point of the sarcoplasm appear to retard protein denaturation as compared with storage at the corresponding temperatures after the greater part of the water has been frozen by exposure to the lower temperatures.

Fish Species and Elapsed Time between Harvest and Freezing. From his data on four kinds of fish, Dyer (*21, 22*) suggested that fatty fishes are more stable in frozen storage than lean fishes. However, more recent data indicate that species differences may sometimes be more important than fat content (*2, 43, 44*). For example, the pattern of ice crystal formation differs between Alaska pollack and yellowtail muscles (*34*) and this may influence stability.

Freshness or post-mortem condition of fish muscle at the time of freezing has an important bearing on the rate of freeze denaturation (*1, 2, 38, 45, 46*).

The stability of fish muscle in frozen storage also varies with the season and other biological conditions such as nutritional status, degree of fatigue, and spawning status (pre, in, or post) (*45, 47–50*).

Denaturation of Proteins

Methods. Early investigations relevant to technological analyses of the freezing operations involved determining the amount of protein extractable in salt solutions, such as 5% NaCl or 0.6*M* KCl. Proteins extracted in this manner were defined as native or undenatured. Then a question arose concerning the mechanism by which denaturation occurs. Unfortunately, denatured proteins are difficult to study because of their insolubility; thus information about the state of proteins in fish had to be gained from the soluble protein fraction.

Actomyosin. SOLUBILITY. Studies have dealt with changes in the solubility of proteins during frozen storage of fish muscle or solutions of isolated actomyosin (*33, 51, 52*). Analysis by gel filtration of the salt extracts has shown that the actomyosin fraction decreases in solubility during frozen storage whereas the sarcoplasmic proteins remain essentially unchanged (*53*).

VISCOSITY. The reduced viscosity of the protein extracted from frozen–stored fish muscle (*44, 54*) and of the soluble fraction of the frozen–stored solutions of isolated actomyosin decreases with increasing time of storage (*51, 52*).

SEDIMENTATION. Examination of extracts from fish muscle or actomyosin solutions by ultracentrifuging procedures has indicated that the actomyosin components (20S–30S) decrease with increasing time of frozen storage and several faster moving components simultaneously appear. After prolonged storage, no components with movement faster than 20S are evident (*29, 51*), suggesting a progressive polymerization of actomyosin. Detailed analyses of the ultracentrifugal patterns of extracts from frozen–stored cod muscle were done by King and his coworkers (*55, 56, 57*) and they reported that G–actomyosin forms as an intermediate product during frozen storage and that some amount of G–actin is eventually dissociated from G–actomyosin. The existence of G–actomyosin is questionable and deserves further study (*2*).

The changes that occur in viscosity and ultracentrifugal pattern during frozen storage suggest deformation of actomyosin filaments into some denser, less asymmetric form.

SALTING-OUT ANALYSIS. Analysis of salting-out curves of extracts from fish muscles before and after frozen storage has shown that frozen storage shifts the actomyosin peak to a higher saturation level (*44, 58, 59*), with variable results for different fish species (*44*). A similar experiment on a solution of isolated carp actomyosin has shown a similar shift of the peak from 35% to 38% saturation ($(NH_4)_2SO_4$) following frozen storage, suggesting the release of some actin from the actomyosin complex (*59*).

ATPASE ACTIVITY AND RELATED PROPERTIES. During frozen storage of cod fillets, a decrease is observed in the ATPase activity of actomyosin extracts, but the number of free SH groups does not change (*32*). This decrease in ATPase activity also has been reported for actomyosin from various other fishes in situ during frozen storage (*60, 61, 62*). During frozen storage of solutions or suspensions of actomyosin isolated from carp, ATPase activity initially rises and then declines to zero (*42, 51, 63, 64*). The initial rise in activity suggests a slight conformational deformation around the active sites followed by conversion to an inactive random state. The decrease in ATPase activity of actomyosin is reflected in the decline in its superprecipitation capacity (*51, 52*) and in the rate of viscosity drop on the addition of ATP (*29, 59*).

ANALYSIS OF SOLUBILITY–STORAGE CURVES. The solubility–storage time curves of actomyosin and myosin extracted from the same frozen cod fillets form a family of parallel lines that are not linear (*65*), whereas the solubility of actin does not change with storage time (*65, 66*). Thus Connell attributed the freeze denaturation or loss of solubility of actomyosin to the myosin component. The author's group determined the amount of soluble protein that remained in a frozen suspension of actomyosin isolated from carp and plotted this value on logarithmic scale against storage time. The plot consisted of two intersecting straight lines with different slopes, suggesting that the denaturation process involved more than one mechanism (*67*). The nature of the plot did not change when the ionic strength of the storage solution was changed but it did change when the pH was altered. Replotting the results of Dyer (*20, 21, 22*) and Connell (*65*) in the manner described yielded plots for fish muscle that were similar to the plot just described.

Analysis of the soluble-protein fractions remaining after frozen storage of fish is being accomplished in the author's laboratory by sodium dodecyl sulfate–polyacrylamide gel electrophoresis (*68*). Tropomyosin remains in solution after long storage and it seems to be the most stable protein during frozen storage. Actin is the next most stable protein. During the early stages of freeze denaturation of proteins in fish both myosin and actin apparently form an insoluble fraction that accounts for the observed decrease in protein solubility.

ELECTRON MICROSCOPY. Examination of fish proteins by electron microscopy conclusively shows that actomyosin aggregates during frozen storage (*59, 63, 69*). The change in structures of the extracted myofibrillar proteins and of the myofibril residues of frozen–stored cod muscle was studied by electron microscopy. The decrease in the number of actomyosin filaments and an increase in the number and size of large aggregate were found (*69*). Unfrozen carp actomyosin, either dissolved in $0.6M$ KCl or suspended in $0.05M$ KCl, exists in a typical arrowhead structure (*59, 63*). After freezing followed by immediate thawing the actomyosin filaments become loosely joined together in a crosswise fashion. After frozen storage for several weeks, the native structure is lost and only masses or aggregates of randomly entangled filaments are evident. Careful examination of the pictures suggests that the aggregated actomyosin filaments are thinner and more flexible than those found in the unfrozen preparation, i.e., the arrowhead structure is no longer apparent. Also evident in the background of the pictures are other particles that may consist partly of aggregated masses of myosin molecules. The thinned entangled filaments in the aggregates are probably filaments of F-actin.

Changes in the shape of carp actomyosin filaments during frozen storage was also studied at various pH's (*67*). Pictures of actomyosin frozen and stored in 0.9*M* and 1.2*M* KCl illustrated the dissociation of actomyosin into actin and myosin (*1, 2*).

Myosin. AGGREGATION. Because of the difficulty of isolating pure myosin from fish studies on the behavior of this protein were delayed. Connell studied the behavior of cod myosin during freezing to temperatures ranging from about -7 to $-78°C$. By ultracentrifugal analysis he showed that the myosin monomer polymerizes to dimers, trimers, and other larger multiples (*70*). Neither the specific rotation nor the number of active SH groups changed appreciably during polymerization. Connell suggested that myosin molecules aggregate side-to-side without unfolding or undergoing any change in intramolecular conformation. The aggregation was ascribed not to S–S bonding but to linkages of an unknown nature. Connell also suggested that myosin has a major role in the insolubilization of actomyosin during frozen storage (*2, 65, 66*).

Changes in the solubility, ultracentrifugal behavior, and number of SH groups in frozen myosin from trout were studied by Buttkus (*71, 72*).

ATPASE. Myosins isolated from various frozen–stored fish muscles exhibit specific activities for ATPase that are slightly lower than the specific activities of myosin from fresh muscles (*73, 74*). The decline of ATPase activity also occurs with myosin isolated from carp, when storage is conducted at $-20°C$ (*64*). Like carp actomyosin, the decline is preceded by a temporary rise in activity.

FILAMENT FORMATION. Like rabbit myosin (*74*), carp myosin in solutions of low ionic strength forms filaments that are observable under the electron microscope. These filaments are either spindle shaped or dumbbell shaped depending upon the method used to prepare thenm. Filaments formed from either frozen–stored samples (filaments or myosin solution) were not as perfect in shape as those prepared from unfrozen, intact myosin. In frozen storage, the spindle-shaped myosin was more stable than the dumbbell-shaped myosin, and myosin in the dissolved state was least able to form filaments (*64*).

Subunits of Myosin. Matsumoto et al. (*64*) isolated H–meromyosin (HMM) and L-meromyosin (LMM) from carp muscle (*15*) and studied their stabilities at −20°C. The ATPase activity of HMM decreased much faster than that of myosin and the capacity of HMM to bind with F–actin as determined by electron microscopy was lost. LMM also exhibited a decreased capacity to form well-ordered paracrystals. These results tend to indicate that frozen storage causes myosin molecules to aggregate side-by-side and myosin subunits to undergo conformational deformations.

Actin. According to Connell (*65, 66*), actin in cod muscle remains essentially unmodified during storage for as long as 30 weeks at −14°C. This statement is based on a study of various physico–chemical properties such as intrinsic viscosity, myosin combining ability, and ultracentrifugal behavior.

On the other hand, actin extracts frozen for either 96 or 127 weeks at −20°C undergo significant alteration. This was shown in the author's laboratory using actin isolated from carp (*75*). For these studies actin was prepared by Guba–Straub's method with the exception that Spudich–Watt's buffer A was used. Actin in either G-form or F-form was frozen and stored at −20°C, and its solubility, reduced viscosity, polymerizing ability (G-actin), and appearance under the electron microscope (F-actin) were tested periodically. Both G- and F-actins underwent denaturation during frozen storage. As indicated earlier, actomyosin, as judged by its electrophoretic (SDS-disc electrophoresis) behavior, becomes increasingly insoluble during frozen storage (*68*).

The disagreement between Connell's results and those by the author's group probably arises because of the different experimental conditions used. In Connell's study, actin was in situ, whereas in the author's study isolated actin or actomyosin was used.

Tropomyosin and Troponin. Tropomyosin is apparently the most stable of the fish proteins during frozen storage. Tropomyosin can be extracted long after actin and myosin become inextractable (*68*).

Troponins isolated from frozen-stored bigeye tuna, *Tilapia* or *Beryx*, are less active in their regulatory function than those from fresh fish muscles (*76*).

Myofibrils and Tissues. Microscopic observations of frozen cod muscle were first published by Love (*23, 24, 25, 26*) and similar observations of other fish species were reported by Tanaka (*27, 34, 48*).

To determine the degree of adherence of myofibrils, Love and MacKay (*77*) developed the "cell fragility method." This method involves measuring the turbidity of thawed muscle homogenates prepared in 1.15% formaldehyde. It has been shown that the extinction value decreases with increasing length of frozen storage, indicating reduced dispersion of the myofibrils. This method has been helpful in studying several problems (*30, 78*).

Electron microscopic study of frozen-stored carp muscle have shown that the ultrastructure of myofibrils is, at least in some specimens, well preserved during storage for 60 weeks at −20°C (*34*). The ultrastructural changes during frozen storage of cod muscle (*79*), cod myofibrils, and solutions of myofibrillar proteins (*80*) have been studied using the freeze etching and the negative staining techniques. Muscle stored for a long period at −20°C showed disturbances in the hexagonal pattern, deformation of the sarcoplasmic reticulum, and a significant reduction in the interfilament spacing. Additional studies of a similar nature have been published recently (*81, 82*).

Connective Tissue Proteins. Collagen comprises the major material of skin and myocommata. Love and his co-workers studied the problem of gaping in which slits or holes appear in the muscle and sometimes the fillet falls apart. This defect is related to the behavior of myocommata proteins (*83–87*). Gaping is more severe in fishes frozen in-rigor than in those frozen pre-rigor, and the seriousness of this defect varies with the biological condition of the fish (influenced by season, size, age, and whether the fish is healthy or starved), and the fish species.

On freezing, ice crystals sometimes form in the myocommata and this may influence the incidence of gaping. During studies where the pH and moisture of the muscle were variables, it was learned that gaping was correlated with the mechanical strength of myocommata collagen that is less strong at lower pH's. Gaping was found to vary with fish species whereas the physico–chemical properties of collagen from various species were found to be identical with normal collagens.

Tanaka (*34*) reported that muscle–cell membranes of Alaska pollack became tougher during frozen storage.

Other Proteins. Since Reay and Dyer discovered that denaturation of myofibrillar proteins is of such profound importance, little attention has been given to the water-soluble proteins including enzymes and other proteins in the sarcoplasm, subcellular organelles, and cell membranes. Recently reports have appeared on the freeze denaturation of enzymes. These studies involved enzymes such as catalase, ADH, GDH, LDH, and MDH from sources other than fish (*88, 89, 90*) and attention was given to the effectiveness of various cryoprotective substances (*89, 90*). Comparable studies with enzymes from fish muscle are few in number (*91*). Studies on fish muscle proteins must be extended to this area if a complete picture of the freeze denaturation of fish muscle is to be obtained. It should be noted that freeze stable enzymes might have important effects during frozen storage of fish (*92, 93*).

Myoglobin undergoes changes during frozen storage and this is the cause of discoloration of red muscle fishes such as tuna. During frozen storage myoglobin is denatured, perhaps via metmyoglobin. Discoloration of tuna myoglobin can be prevented by rapid freezing and storage at about −60°C (*94*), conditions that are employed commercially in Japan. By freezing with liquid nitrogen discoloration of tuna myoglobin is reduced (*95*).

Cause of Denaturation

According to Love (*41*) freezing itself has a detrimental effect on proteins. Some aggregation of actomyosin filaments has been observed in specimens thawed immediately after freezing to −20°C (*59*).

Several proposals have been advanced concerning the causes and mechanisms for freeze denaturation of fish muscle proteins but the views usually have some features in common. In all instances, the formation of ice crystals has effects, either directly or indirectly, that are of major importance.

Effect of Mineral Salts. As cellular water is frozen, mineral salts and soluble–organic substances become concentrated in the remaining unfrozen phase. This increase in solute concentration, with corresponding changes in ionic strength and pH, is believed to affect dissociation and/or denaturation of proteins (*1, 2, 62, 96–98*). Experiments by Fukumi et al. (*99*) support this theory. They found that freeze denaturation of washed actomyosin from Alaska pollack muscle was accelerated by the presence of Ca^{2+}, Mg^{2+}, K^{+}, and Na^{+} ions and reduced by their removal.

Dehydration of Proteins. Ice crystal formation, especially when brought about by slow freezing, results in a redistribution of water. When a cell is thawed water has difficulty returning to its original sites (e.g., associated with proteins) and some of this water leaves the tissue (drip). Denaturation or association of protein molecules that may occur during freezing and frozen storage hinders the process of protein rehydration (*1, 2, 4*).

As ice forms and water molecules are removed from areas near the proteins, the protein molecules move closer together and the possibility of aggregation through intermolecular cross-bonding increases. If freezing is carried out to a very low temperature, removal of water from proteins may occur to such an extent that the proteins might be affected.

Evidence that aggregation occurs during frozen storage is provided in the form of physico–chemical data presented earlier. Evidence for freeze-induced changes in conformation is provided from the changes in biochemical properties of myosin and its subunits (e.g., changes in ATPase activity, superprecipitation capacity, actin-binding capacity, and ability to form paracrystals) as described earlier. The effects of cryoprotective substances, as mentioned later, provide further support for these views.

Based on the decrease in active SH groups that occurs in trout myosin during frozen storage, Buttkus (*71, 72*) proposed the formation of intermolecular S–S bonds as a major cause of aggregation. Myosin aggregates were soluble in 6*M* guanidine–HCl containing 0.5*M* mercaptoethanol but were not soluble in 1*M* NaCl, 8*M* urea, 6*M* guanidine–HCl, or detergents. Hydrophobic and hydrogen bonds were suggested also as being involved in the aggregation process. The proposed mechanism of aggregation consisted mainly of S–S bond rearrangement and an influence of KCl at its eutectic point.

Findings in the author's laboratory (*100*) demonstrated that the number of cross bonds in carp actomyosin and myosin increases during frozen storage and that solubility of these proteins decreases. Based on the types of chemicals that resolubilized these proteins at various rates, it was concluded that ionic bonds, hydrogen bonds, covalent bonds (S–S), and hydrophobic associations all are involved in the aggregation process.

Effects of Lipids. Dyer was the first to suggest that lipids and their derivatives might be involved in protein denaturation during the frozen storage of fish (*21, 22*). Since his studies many workers have investigated this possibility (*101, 102, 103*). A series of elaborate works were carried out by King, Anderson, and their collaborators to determine whether protein–lipid interactions are the cause of protein insolubilization in fish muscle during frozen storage (*55, 56, 57, 104, 105*).

Effects of linoleic acid and linoleic acid hydroperoxides on the myofibrils and the solutions of myofibrillar proteins of cod muscle have been proved using the electron microscopy (*80*). Linoleic acid hydroperoxides were ten times more effective than linoleic acid in reducing the amount of the protein in KCl-extracts from the myofibrils incubated with the acid or its hydroperoxides. Linoleic acid seemed to prevent the dissolution of the myofibril frame work but appeared not to impair the extraction of myosin while hydroperoxides appeared to cause a retention of A-bands (myosin) in the myofibrils.

Nevertheless after reviewing the literature, Connell concluded that protein–lipid interactions do not appear to be a major cause of protein denaturation during frozen storage of fish (*2*).

Effect of Other Substances. GAS. Oxygen dissolved in the blood and internal media of fish may encourage proteins to form S–S bonds during frozen storage. Furthermore, because of its low solubility in ice, nitrogen may be expelled from the frozen medium, resulting in fine bubble formation and surface denaturation of proteins (*27*). This was postulated by Tanaka (*27*) because he observed considerable molecular nitrogen in the holes of so called "sponge meat" of cod that had been frozen for a long time.

HEAVY METALS. Heavy metals are known to denature proteins. Acceleration of freeze denaturation of trout myosin by Cu^{2+} has been observed (*72*). In stored Gadoid muscle, Cu^{2+} accelerates the transformation of trimethylamine oxide to HCHO via trimethylamine and dimethylamine, in turn, and the HCHO might bind to proteins, resulting in their denaturation (*106*).

WATER SOLUBLE PROTEINS. In studies with frozen minced Alaska pollack, the possibility has been raised that removal of water soluble proteins may render the residual actomyosin more stable to frozen storage (*107*). Whether the effect of the washing is because of the removal of the water–soluble proteins or of the organic substances of lower molecular weight is left to further study.

Control of Denaturation

Because protein denaturation usually does not involve a single specific reaction, control of this occurrence at first was thought to be very difficult. Early attempts to reduce denaturation depended almost entirely on control of temperature during frozen storage and on control of fish properties (freshness, biological condition, etc.) prior to freezing. Now, control of freeze denaturation is being approached through the addition of cryoprotective substance and this approach is likely to increase in importance.

Frozen Mince of Alaska Pollack Muscle. In 1959 researchers at Hokkaido Fisheries Research Station, Japan, headed by Nishiya, developed a technique by which freeze denaturation of proteins in Alaska pollack muscle could be prevented (*108, 109*). The technique consists of: 1) removing the mineral salts and water-soluble organic substances from minced muscle by washing with water and 2) addition of 10% sucrose and 0.2–0.5% polyphosphates or a mixture of 5% sucrose, 5% sorbitol and polyphosphate prior to freezing. By this means, actomyosin of Alaska pollack, which is remarkably unstable in frozen storage, is protected from denaturation. Within a few years, this new technique stimulated growth of an industry that now has an annual production of 400,000 tons. Scientific analyses followed the initial invention and these proved the benefits of removing the minerals and water-soluble organic substances (including water-soluble proteins) and of adding the cryoprotective chemicals (*99*).

Cryoprotective Compounds. Cryoprotective effects of sugars, polyalcohols, and compounds of other families have been known since the 1940s. Many works on those effects have been reported in the fields of freezing preservation of blood, microorganisms, etc., but this review will deal with only the works on fish muscle proteins.

Polyphosphates. Treating round fish or fillets with phosphate solutions prior to freezing has proven to have a protective effect on proteins (*97, 110, 111, 112, 113*). Triphosphate and hexametaphosphates are more effective than pyrophosphate and orthophosphate is of little benefit.

Additional studies have been conducted on frozen samples of actomyosin isolated from carp. Actomyosin was either suspended in 0.05*M* KCl or dissolved in 0.6*M* KCl. These studies showed that triphosphate is more cryoprotective than pyrophosphate when actomyosin is suspended in 0.05*M* KCl, whereas both chemicals are ineffective when actomyosin is dissolved in 0.6*M* KCl. Orthophosphate was ineffective in both situations (*52*).

In the frozen mince of Alaska pollack there was a marked synergistic effect between pyrophosphates or polyphosphates and sucrose or sorbitol, but this effect was less significant if 2.5% NaCl was present in the mince (*97, 110, 114*). Also for carp actomyosin solution in 0.6*M* KCl, no appreciable synergistic effect was found among these additives, while some synergistic effect was found in 0.05*M* KCl. When a suspension of carp myofibrils was used instead of the actomyosin preparation, the synergistic effect was greater than with the isolated protein (*115, 116, 117*). This suggests that the synergistic cryoprotective effect of these substances is dependent on the mode of association of actin and myosin and the ionic state of these proteins.

Organic Compounds. Sucrose, glucose, other sugars, and sorbitol have cryoprotective effects on frozen mince of Alaska pollack (*60, 118*) and carp actomyosin (*117, 119, 120*). Furthermore, ethylene glycol, glycerol (*121, 122*), and citrate (*123*) have cryoprotective effects on the proteins in muscles of Alaska pollack and cod.

In the author's laboratory, an in vitro test was devised to evaluate the cryoprotective effect of any compound on fish protein. The test system consisted of carp actomyosin either in solution (in 0.6*M* KCl) or in suspension (in 0.05*M* KCl). By means of this system about 150 compounds were screened, out of which about 30 compounds were found to be markedly effective and another 20 compounds were found to be moderately effective. Among the former group, monosodium glutamate (MSG) was particularly outstanding. Additives were used at a concentration of 0.1–0.2*M* in the final mixture and the pH was adjusted to 7 before freezing (*51, 52, 116, 117, 119, 124, 125, 126*). Similar studies have been done by other workers (*60, 127*).

The effective compounds found thus far are distributed over several classes of chemicals, namely amino acids, dicarboxylic acids, hydrocarboxylic acids, polyalcohols, carbohydrates, and polyphosphates. However, not all members of each class are effective. Among the amino acids none of those with large nonpolar groups, such as valine and leucine are effective. Furthermore, monocarboxylic acids are ineffective. With carbohydrates, effectiveness is present only in those that are smaller than inulin. Thus starch is ineffective unless hydrolyzed into smaller fragments. Differences are also found in the effectiveness of various isomers of monosaccharides (*52, 117, 119*).

Similar studies have been conducted using isolated carp myosin as a test system and the results were essentially the same as described above (*128*).

Mechanism of the Cryoprotective Effect. The results on all of the compounds tested were systematically evaluated and an attempt was made to correlate molecular structure of the test compound with its cryoprotective effectiveness. This evaluation led to a proposed list of chemical attributes that seem to be characteristic of cryoprotective substances: 1) molecule has to possess one essential group, either $-COOH$, $-OH$, or $-OPO_3H_2$, and more than one supplementary group, $-COOH$, $-OH$, $-NH_2$, $-SH$, $-SO_3H$, and/or $-OPO_3H_2$; 2) the functional groups must be suitably spaced and properly oriented with each other and 3) the molecule must be comparatively small (*52, 117*).

By taking these requirements into account, a mechanism for the cryoprotective effect was proposed. Each cryoprotective compound appears to function as a coating material by associating with the protein by ionic bonding or hydrogen bonding, the means depending on the nature of the compound. The ionic coating presumably occurs with acidic or basic amino acids and with dicarboxylic acids. In these instances the compound associates through their ionic groups with the oppositely charged sites of proteins, thereby increasing the net charge of protein, increasing its electrostatic repulsion, and hindering aggregation of protein molecules during frozen storage. The increased net charge also may augment protein hydration.

On the other hand, the hydrogen-bond coating appears to be operative when carbohydrates and polyalcohols are used. In these instances, the added compound presumably covers the protein molecules by hydrogen bonding with the OH groups of the protein. The "extra" OH groups of the additive molecules would hydrogen bond with water, thereby increasing hydration of the protein molecules and hindering their aggregation (*52, 117*).

An alternate mechanism has recently been considered for carbohydrates and polyalcohols (*129, 130*). These additives might alter the state of liquid water so as to impede ice–crystal formation. This view finds some support from data obtained by differential thermal analysis at low temperatures (*130*).

Cysteine apparently functions by coating the protein molecules through S–S bonding because this compound supplements the effect of MSG (*116, 117*).

Sodium Glutamate and Its Effect on Muscle Proteins. MSG is a highly effective compound for protecting carp actomyosin against freeze denaturation and it functions at a concentration as low as 0.025*M*. When carp actomyosin is stored frozen in the presence of 0.1–0.2*M* glutamate, little or no denaturation occurs during 10 weeks storage at −20 to −30°C (denaturation was monitored by measuring various physico–chemical and biochemical properties of the protein samples). It is particularly remarkable that samples of actomyosin, when stored in the presence of MSG, exhibit little change in appearance (electron microscope) during frozen storage (*51, 63, 131*).

MSG is also an effective cryoprotective agent for myosin, HMM, LMM, and actin that have been isolated from carp muscle (*63, 64*). Collectively, these results suggest that the cryoprotective effect of MSG extends to each constituent protein and each subunit of the myofibrils.

Electron microscopy studies of actomyosin that have been frozen and stored in the presence of 1*M* glucose indicate that glucose is essentially as effective as MSG (*132*).

Summary

Now it can be said that the nature and mechanisms of protein denaturation in frozen fish muscle are becoming clearer. Denaturation is evident not only at the level of highly organized whole muscle but also at the level of less organized intracellular constituents involving associated protein systems, individual protein molecules, and subunits of proteins. Detailed information on changes in intramolecular conformation during freezing is still lacking and is needed. Also needed is further information on the effect of freezing on intracellular organelles.

Controlling freeze denaturation, which appeared very difficult only a short time ago, is now within reach as far as minced muscle is concerned. MSG has proved to be an especially effective cryoprotective agent. Still to be accomplished is the control of protein denaturation in intact fish muscle.

Literature Cited

1. Dyer, W. J.; Dingle, J. R. In "Fish as Food"; Borgstrom, G., Ed.; Academic: New York, 1961; Vol. 1, pp 275–327.
2. Connell, J. J. In "Low Temperature Biology of Foods"; Hawthorn, J., Rolfe, E. J., Eds.; Pergamon: Oxford, 1968; pp 333–358.
3. Love, R. M. In "Cryobiology"; Meryman, H. T., Ed.; Academic: London, 1966; pp 317–405.

4. Matsumoto, J. J. *Reito (Refrigeration)* **1972**, *47*, 206–215.
5. Powrie, W. O. In "Low Temperature Preservation of Foods and Living Matter"; Marcel Dekker: New York, 1973.
6. Fennema, O. R. In "Low Temperature Preservation of Foods and Living Matter"; Marcel Dekker: New York, 1973.
7. Sikorski, Z.; Olley, J.; Kostuch, S. In "Critical Reviews in Food Science and Nutrition"; CRC: Cleveland, 1976; pp 97–129.
8. Noguchi, S. In "Proteins of Fish"; Jpn. Soc. Sci. Fish., Ed.; Koseisha–Koseikaku, K. K.: Tokyo, 1977; pp 91–108.
9. Love, R. M. "Chemical Biology of Fishes"; Academic: New York, 1970; pp 547.
10. Huxley, H. E. In "The Structure and Function of Muscle"; Bourne, G. H., Ed.; Academic: New York, 1972; Vol. 1, Part A, pp 301–387.
11. Hamoir, G. In "Advances in Protein Chemistry"; Academic: New York, 1955; Vol. 10, pp 227–288.
12. Jaquot, R. In "Fish as Foods"; Borgstrom, G., Ed.; Academic: New York, Vol. 1, pp 145–209.
13. Seki, N. In "Proteins of Fish"; Jpn. Soc. Sci. Fish., Ed.; Koseisha–Koseikaku, K. K.: Tokyo, 1977; pp 7–42.
14. Connell, J. J. *Biochem. J.* **1959**, *71*, 83–86.
15. Tsuchiya, T.; Matsumoto, J. J. *Bull. Jpn. Soc. Sci. Fish.* **1975**, *41*, 1319–1326.
16. Connell, J. J. *Biochem. J.* **1960**, *75*, 530–538.
17. Connell, J. J. *Biochem. J.* **1961**, *80*, 503–509.
18. Finn, D. B. *Proc. R. Soc. (London)* **1932**, *B111*, 396–411.
19. Reay, G. A.; Kuchel, C. C. *Gr. Brit. Dept. Sci. Ind. Research Rep. Food Invest. Board* **1937**, *1936*, 93–95.
20. Dyer, W. J. *Food Res.* **1951**, *16*, 522–527.
21. Dyer, W. J.; Morton, M. L. *J. Fish. Res. Board Can.* **1956**, *13*, 129–134.
22. Dyer, W. J.; Morton, M. L.; Fraser, D. I.; Bligh, E. G. *J. Fish. Res. Board Can.* **1956**, *13*, 569–579.
23. Love, R. M. *J. Sci. Food Agric.* **1955**, *6*, 30–37.
24. Love, R. M.; Karsti, O. *J. Sci. Food Agric.* **1958**, *9*, 249–256.
25. Love, R. M. *J. Sci. Food Agric.* **1958**, *9*, 262–268.
26. Love, R. M. *J. Sci. Food Agric.* **1958**, *9*, 609–617.
27. Tanaka, T. In "Water in Food"; Jpn. Soc. Sci. Fish., Ed., Koseisha–Koseikaku K. K.: Tokyo, 1973; pp 63–82.
28. Matsuda, Y. *Bull. Jpn. Soc. Sci. Fish.* **1969**, *35*, 891–896.
29. Suzuki, T.; Kanna, K.; Tanaka, T. *Bull. Jpn. Soc. Sci. Fish.* **1964**, *30*, 1022–1036.
30. Love, R. M. *Bull. Jpn. Soc. Sci. Fish.* **1967**, *33*, 746–752.
31. Love, R. M. *Chem. Ind.* **1967**, 2151.
32. Connell, J. J. *J. Sci. Food Agric.* **1960**, *11*, 245–249.
33. Snow, J. M. *J. Fish. Res. Board Can.* **1950**, *7*, 599–607.
34. Tanaka, T. In "The Technology of Fish Utilization"; Kreuzer, R., Ed.; Fishing News (Books) Inc.: London, 1965; pp 121–125.
35. Love, R. M. *J. Sci. Food Agric.* **1966**, *17*, 465–471.
36. Dyer, W. J.; Fraser, D. I.; Ellis, D. G.; MacCallum, W. A. *J. Fish. Res. Board Can.* **1957**, *14*, 627–635.
37. Dyer, W. J.; Fraser, D. I.; MacCallum, W. A. *J. Fish. Res. Board Can.* **1957**, *14*, 925–929.
38. Nikkilä, O. E.; Linko, R. R. *Food Res.* **1954**, *19*, 200–205.
39. Dyer, W. J.; Fraser, D. I.; Ellis, D. G.; Idler, D. R.; MacCallum, W. A.; Laishley, E. *Bull. Inst. Refrig.* **1962**, Supplement.
40. Dyer, W. J.; Fraser, D. I.; Jewell, G. J.; James, L. *Progress Rep. Atlantic Coast Station, Fish. Res. Board Can.* **1962**, *73*, 7–20.
41. Love, R. M. *Nature* **1964**, *211*, 981–982.

42. Kato, N.; Umemoto, S.; Uchiyama, H. *Bull. Jpn. Soc. Sci. Fish.* **1974,** *40,* 1263–1267.
43. Nikkilä, O. E.; Linko, R. R. *Food Res.* **1956,** *21,* 42–46.
44. Ueda, T.; Shimizu, Y.; Simidu, W. *Bull. Jpn. Soc. Sci. Fish.* **1962,** *28,* 1005–1009, 1010–1014.
45. Fukuda, Y.; Kakehata, K.; Arai, K. *Bull. Jpn. Soc. Sci. Fish.* **1977,** *43,* 715–725.
46. Love, R. M. *J. Sci. Food Agric.* **1962,** *13,* 534–545.
47. Fraser, D. I.; Dyer, W. J.; Weinstein, H. M.; Dingle, J. R.; Hines, J. A. *J. Fish. Res. Board Can.* **1966,** *44,* 1015–1033.
48. Tanaka, T. *Bull. Tokai Reg. Fish. Res. Lab.* **1969,** *60,* 143–168.
49. Umemoto, S.; Kanna, K. *Bull. Tokai Reg. Fish. Res. Lab.* **1969,** *60,* 169–177.
50. Love, R. M.; Haraldsson, S. B. *J. Food Sci.* **1961,** *12,* 442–449.
51. Noguchi, S.; Matsumoto, J. J. *Bull. Jpn. Soc. Sci. Fish.* **1970,** *36,* 1078–1087.
52. Matsumoto, J. J.; Noguchi, S. *Proc. Int. Congr. Refrig., 13th, 1971,* Vol. 3, 237–241.
53. Umemoto, S.; Kanna, K.; Iwata, K. *Bull. Jpn. Soc. Sci. Fish.* **1971,** *37,* 1100–1104.
54. Seagran, H. L. *Food Res.* **1956,** *23,* 143–149.
55. King, F. J. *J. Food Sci.* **1966,** *31,* 649–663.
56. Anderson, M. L.; Ravesi, E. M. *J. Fish. Res. Board Can.* **1969,** *26,* 2727–2736.
57. Anderson, M. L.; Ravesi, E. M. *J. Food Sci.* **1970,** *35,* 199–206.
58. Migita, M.; Otake, S. *Bull. Jpn. Soc. Sci. Fish.* **1961,** *27,* 327–338.
59. Oguni, M.; Kubo, T.; Matsumoto, J. J. *Bull. Jpn. Soc. Sci. Fish.* **1975,** *41,* 1113–1123.
60. Arai, K.; Takashi, R. *Bull. Jpn. Soc. Sci. Fish.* **1973,** *39,* 533–541.
61. Arai, K.; Kawamura, K.; Hayashi, C. *Bull. Jpn. Soc. Sci. Fish.* **1973,** *39,* 1077–1085.
62. Ohta, F.; Yamada, T. *Bull. Jpn. Soc. Sci. Fish.* **1978,** *44,* 63–66.
63. Tsuchiya, T.; Tsuchiya, Y.; Nonomura, Y.; Matsumoto, J. J. *J. Biochem.* **1975,** *77,* 853–862.
64. Matsumoto, J. J.; Tsuchiya, T.; Noguchi, S.; Ohnishi, M.; Akahane, T.; published at 26th Int. Congr. Pure Appl. Chem., Tokyo, Sept. 8, 1977.
65. Connell, J. J. *J. Food Sci. Agric.* **1962,** *13,* 607–617.
66. Connell, J. J. *J. Food Sci. Agric.* **1960,** *11,* 515–519.
67. Ohnishi, M.; Tsuchiya, T.; Matsumoto, J. J. *Bull. Jpn. Soc. Sci. Fish.* **1978,** *44,* 27–37.
68. Irisa, Y.; Ohnishi, M.; Tsuchiya, T.; Matsumoto, J. J.; unpublished data.
69. Jarenbäck, L.; Liljemark, A. *J. Food Technol.* **1975,** *10,* 309–325.
70. Connell, J. J. *Nature* **1959,** *183,* 664–665.
71. Buttkus, H. *J. Food Sci.* **1970,** *35,* 558–562.
72. Buttkus, H. *Can. J. Biochem.* **1971,** *49,* 97–107.
73. Kimura, I.; Murozuka, T.; Arai, K. *Bull. Jpn. Soc. Sci. Fish.* **1977,** *43,* 315–321.
74. Kimura, I.; Murozuka, T.; Arai, K. *Bull. Jpn. Soc. Sci. Fish.* **1977,** *43,* 795–803.
75. Tsuchiya, T.; Nakamura, Y.; Ohnishi, M.; Matsumoto, J. J.; published at Annual Meeting Jpn. Soc. Sci. Fish., Tokyo, April 2, 1977.
76. Seki, N.; Hasegawa, E. *Bull. Jpn. Soc. Sci. Fish.* **1978,** *44,* 71–75.
77. Love, R. M.; MacKay, E. M. *J. Sci. Food Agric.* **1962,** *13,* 200–212.
78. Love, R. M.; Aref, M. M.; Elerian, M. K.; Ironside, J. I. M.; MacKay, E. M.; Varela, M. G. *J. Sci. Food Agric.* **1965,** *16,* 259–267.
79. Jarenbäck, L.; Liljemark, A. *J. Food Technol.* **1975,** *10,* 229–239.
80. Jarenbäck, L.; Liljemark, A. *J. Food Technol.* **1975,** *10,* 437–452.

81. Ohnishi, M.; Tsuchiya, T.; Matsumoto, J. J.; published at Annual Meeting Jpn. Soc. Sci. Fish., Tokyo, April 2, 1978.
82. Kojima, N.; Yamada, T.; Oosato, T.; Inoue, Y.; published at Meeting Jpn. Soc. Sci. Fish., Sendai, Oct. 2, 1977.
83. Love, R. M.; Robertson, I. *J. Food Technol.* **1968,** *3,* 215–221.
84. Love, R. M.; Lavety, J.; Garcia, N. G. *J. Food Technol.* **1972,** *7,* 291–301.
85. Love, R. M.; Lavety, J. *J. Food Technol.* **1972,** *7,* 431–441.
86. Yamaguchi, K.; Lavety, J.; Love, R. M. *J. Food Technol.* **1976,** *11,* 389–399.
87. Love, R. M.; Yamaguchi, K.; Creac'h, Y.; Lavety, J. *Comp. Biochem. Physiol.* **1976,** *55B,* 487–492.
88. Tappel, A. L. In "Cryobiology"; Meryman, H. T., Ed.; Academic: London, 1966; pp 163–177.
89. Hanafusa, N. In "Water in Food"; Jpn. Soc. Sci. Fish., Ed.; Koseisha-Koseikaku K. K.: Tokyo, 1973; pp 9–24.
90. Sakuma, R.; Aoyama, T.; Akahane, T.; Tamiya, T.; Tsuchiya ,T.; Matsumoto, J. J.; unpublished data.
91. Yamanaka, H.; MacKie, I. M. *Bull. Jpn. Soc. Sci. Fish.* **1971,** *37,* 1105–1109.
92. Wood, J. P. *Can. J. Biochem.* **1959,** *37,* 937.
93. Tomlinson, N. *J. Fish. Res. Board Can.* **1963,** *20,* 1145.
94. Bito, M. *Bull. Jpn. Soc. Sci. Fish.* **1969,** *35,* 1193–1200.
95. Matsumoto, J. J.; Matsuda, E. *Bull. Jpn. Soc. Sci. Fish.* **1967,** *33,* 224–228.
96. Duerr, J. D.; Dyer, W. J. *J. Fish. Res. Board Can.* **1952,** *8,* 325–331.
97. Linko, R. R.; Nikkilä, O. E. *J. Food Sci.* **1961,** *26,* 606–610.
98. Ohta, F.; Tanaka, K. *Bull. Jpn. Soc. Sci. Fish.* **1978,** *44,* 59–62.
99. Fukumi, T.; Tamoto, K.; Hidesato, T. *Mon. Rep. Hokkaido Municipal Fish. Exp. Stn.* **1965,** *22,* 30–38.
100. Tsuchiya, Y.; Deura, K.; Tsuchiya, T.; Matsumoto, J. J.; published at Annual Meeting Jpn. Soc. Sci. Fish., Tokyo, April 2 ,1975.
101. Dyer, W. J.; Fraser, D. I. *J. Fish. Res. Board Can.* **1959,** *16,* 43–52.
102. Bligh, E. G. *J. Fish. Res. Board Can.* **1961,** *18,* 143.
103. Hanson, S. W. F.; Olley, J. In "Technology of Fish Utilization"; Kreuzer, R., Ed.; Fishing News (Books) Ltd.: London, 1965; pp 111–115.
104. Anderson, M. L.; Ravesi, E. M. *J. Fish. Res. Board Can.* **1958,** *25,* 2059–2069.
105. Anderson, M. L.; Ravesi, E. M. *J. Food Sci.* **1970,** *35,* 551–558.
106. Dyer, W. J. *Reito (Refrigeration)* **1973,** *48,* 38–41.
107. Tamoto, K. *New Foods Industry* **1971,** *13*(12), 61–69.
108. Nishiya, K.; Takeda, F.; Tamoto, K.; Tanaka, O.; Fukumi, T.; Kitabayashi, T.; Aizawa, S. *Mon. Rep. Hokkaido Municipal Fish. Exp. Stn.* **1961,** *18,* 122–135.
109. Nishiya, K.; Takeda, F.; Tamoto, K.; Tanaka, O.; Kitabayashi, T. *Mon. Rep. Hokkaido Municipal Fish. Exp. Stn.* **1961,** *18,* 391–397.
110. Tamoto, K.; Hidesato, T.; Kida, K. *Mon. Rep. Hokkaido Municipal Fish. Exp. Stn.* **1971,** *28,* 18–38.
111. Tanikawa, E.; Akiba, M.; Shitamori, A. *Food Technol.* **1963,** *17,* 87.
112. Love, R. M.; Abel, G. *J. Food Technol.* **1966,** *1,* 323–333.
113. Dyer, W. J.; Brockenhoff, H.; Hoyle, R. J.; Fraser, D. I. *J. Fish. Res. Board Can.* **1964,** *21,* 101–106.
114. Tamoto, K.; Hidesato, T.; Fukumi, T. *Bull. Hokkaido Reg. Fish. Res. Lab.* **1965,** *22,* 165–170.
115. Kawashima, T.; Arai, K.; Saito, T. *Bull. Jpn. Soc. Sci. Fish.* **1973,** *39,* 525–532.
116. Noguchi, S.; Matsumoto, J. J. *Bull. Jpn. Soc. Sci. Fish.* **1971,** *37,* 1115–1122.

117. Noguchi, S. Doctoral Thesis, Sophia University, 1974.
118. Tamoto, K.; Tanaka, O.; Takeda, F.; Fukumi, T.; Nishiya, K. *Bull. Hokkaido Reg. Fish. Res. Lab., Fish. Agency* **1961,** *23,* 50–60.
119. Noguchi, S.; Oosawa, K.; Matsumoto, J. J. *Bull. Jpn. Soc. Sci. Fish.* **1976,** *42,* 77–82.
120. Arai, K.; Takashi, R.; Saito, T. *Bull. Jpn. Soc. Sci. Fish.* **1970,** *36,* 232–236.
121. Love, R. M.; Elerian, M. K. *J. Sci. Food Agric.* **1965,** *16,* 65–70.
122. Tamoto, K. *Sci. Rep. Hokkaido Fish. Exp. Stn.* **1963,** *1,* 20–25.
123. Okada, M. *Reito (Refrigeration)* **1968,** *42,* 71–81.
124. Noguchi, S.; Matsumoto, J. J. *Bull. Jpn. Soc. Sci. Fish.* **1975,** *41,* 243–249.
125. Noguchi, S.; Matsumoto, J. J. *Bull. Jpn. Soc. Sci. Fish.* **1975,** *41,* 329–335.
126. Noguchi, S.; Shinoda, E.; Matsumoto, J. J. *Bull. Jpn. Soc. Sci. Fish.* **1975,** *41,* 776–786.
127. Nakamura, M. *Mon. Rep. Hokkaido Municpl. Fish. Exp. Stn.* **1970,** *27,* 127–132.
128. Tsuchiya, Y.; Tsuchiya, T.; Deura, K.; Matsumoto, J. J.; unpublished data.
129. Noguchi, S., personal communication.
130. Love, R. M., personal communication.
131. Ohnishi, M.; Tsuchiya, T.; Matsumoto, J. J. *Bull. Jpn. Soc. Sci. Fish.* **1978,** *44,* 755–762.
132. Tsuchiya, Y.; Tsuchiya, T.; Matsumoto, J. J.; unpublished data.

RECEIVED June 16, 1978.

INDEX

C

D

E

F

G

N

O

P

R

The text of this book is set in 10 point Caledonia with two points of leading. The chapter numerals are set in 30 point Garamond; the chapter titles are set in 18 point Garamond Bold.

The book is printed offset on Text White Opaque 50-pound
The cover is Joanna Book Binding blue linen.

Jacket design by Carol Conway.

The book was composed by Service Composition Co., Baltimore, MD, printed and bound by The Maple Press Co., York, PA.